CRITICAL PATH ANALYSIS
IN PRACTICE

Contributors

D. J. ARMSTRONG   *formerly of* Richard Costain

E. L. BUESNEL   Unilever Limited

J. A. CARRUTHERS   Unilever Limited, *formerly of* Richard Costain, *Chairman Critical Path Analysis Study Group 1964–66*

R. COKER   British Ship Research Association, *formerly of* W. S. Atkins & Partners
Chairman Critical Path Analysis Study Group 1966–67
Secretary 1965–66

D. W. FORSYTH   P-E Consulting Group Limited

D. J. MCLEOD   Hunting Engineering Limited

C. STAFFURTH   Hunting Engineering Limited

R. C. J. TAYLOR   Central Electricity Generating Board
Secretary Critical Path Analysis Group Study 1966–67

E. G. TRIMBLE   Department of Civil Engineering, Loughborough University of Technology, *formerly of* the National Building Agency

P. M. USHER   J. & E. Hall Limited
formerly of P-E Consulting Group Limited

S. VAJDA   Department of Engineering Production University of Birmingham

H. WALTON   The British Oxygen Company Limited

D. WILLIAMS   Management Consultant

Editor

GAIL THORNLEY   Department of Operational Research, University of Lancaster

Critical Path Analysis in Practice

COLLECTED PAPERS
ON PROJECT CONTROL

Edited by GAIL THORNLEY

Tavistock Publications

LONDON · NEW YORK · SYDNEY · TORONTO · WELLINGTON

First published in 1968
by Tavistock Publications Limited
11 New Fetter Lane, London E.C.4
and printed in Great Britain
in 10 on 12 point Monotype Times
by Butler & Tanner Ltd, Frome and London
© *The Operational Research Society Limited 1968*

The book is published by
Tavistock Publications
under the auspices of
the Operational Research Society

Previously published, under the same auspices:
'Operational Research and the Social Sciences'
edited by J. R. Lawrence, *1966*

Distributed in the U.S.A.
by Barnes & Noble, Inc.

Contents

Appendices

Editor's Note

My grateful thanks are due to the editorial committee, led by Dr J. A. Carruthers, which had already solicited and assembled the papers before handing over the editorial task to me.

With the exception of Chapter 7, I have made only minor alterations necessary for consistency between independently written sections. To give continuity I have grouped the chapters in five Parts, each headed by a brief commentary.

V. G. T.

Introduction

This volume of papers is aimed at the wide gap that exists between the theory and the practice of project control by network methods. It contains some of the papers and reports presented at meetings of the Critical Path Analysis Study Group of the Operational Research Society since its foundation in 1963, together with material specially prepared by a number of study group members who have been applying the methods for several years. The papers are largely a summary of the practical knowledge and expertise of the group. The writers have shared experience in a very wide range of industries, and the principles outlined should similarly have wide applicability. The coverage of the various aspects is not intended to be exhaustive or balanced, but simply to reflect the type and range of problems encountered in practice.

Network Techniques have been outstandingly successful within their relatively brief history of application, and it takes only a very short time to understand their fundamentals. However, these two factors can make the techniques appear deceptively simple to the newcomer, who may find that between theory and successful application lies a gap which it takes considerable effort and skill to bridge. The technique-oriented man will come up against management problems in the practical situation, and the manager approaching the technique will discover many problems in application; many problems appear mere details, but such details can make or mar a successful application. The present volume takes project control as its basic problem, and considers the several aspects to be taken into account in starting and managing a project control system by network methods.

The many acronyms used to describe the techniques have caused some confusion. In these papers Critical Path Analysis (CPA), or Network Analysis, is used as a general name for all planning techniques based on a network diagram. Program Evaluation and Review Technique (PERT), familiar to readers of American literature as a similar generic title, is used solely for the probabilistic time networks described in the original PERT papers.

It is assumed that the reader knows the elements of network methods and now wishes to apply them in practice. The first Part sketches the scope of the techniques and current fields of application in Britain. For completeness, a brief description of the basic methods is given, with a short description of an alternative, the Method of Potentials. Problems of implementation are discussed in Part II; papers cover staffing requirements and choice of the first project, the level of detail desirable in a network diagram, and the use of computers.

In Part III the more complex aspects of the techniques are discussed. Resource methods are a very important aspect, yet rigorous, practical, and generally applicable methods have not yet been developed. The heuristic methods currently available are described and their use is discussed. Two further papers deal with probabilistic techniques: the PERT type probabilistic duration problem, and the problems arising from uncertain logic. The final paper in this Part describes a network-based method of project cost control.

Very large networks bring special problems particularly in the areas of organization and communication. Part IV, based on the report of a Working Party on Large Networks, considers these problems and describes solutions that have been found satisfactory in practical situations.

Various alternative approaches can be made to problems usually solved by network methods; Part V describes a linear programming solution. This approach is not an efficient or recommended method; however, it throws some light on problems of resource allocation and cost minimization. A transportation formulation of CPA is also described.

References within the text are grouped at the end of each paper. The volume is completed by three appendices. The first is a glossary of standard terms and symbols, prepared by members of the study group, which has been submitted to the British Standards Institute as a minimal list. There follows a selective reading list of papers covering the spectrum of network methods and applications most efficiently. Appendix 3 is a table of computer programs available in the UK in the autumn of 1967.

J. A. CARRUTHERS
GAIL THORNLEY

Part I

Review of Basic CPA Methods

The first chapters of this Part outline the problem areas and industries in which CPA has been successfully applied in the UK, and briefly introduce the essential elements of the technique. Newcomers to the subject are advised to refer to a more complete description of the technique; for example in one of the books recommended in the selective reading list.

The arrow-diagram method, described in Chapter 2, is the most commonly used, but is not the only method of drawing network diagrams. Other methods include the circle and link technique which was described at the CPA Study Group meeting of 15 September 1966 by Dr C. J. Anson of Urwick Orr & Partners Ltd, and the method of potentials, which is described in Chapter 3. The different methods have their own advantages and disadvantages, but all are essentially methods of representing the logical relationships between jobs that have to be performed and calculating the consequential time limitations.

1 · The Scope of the Method

D. WILLIAMS

History

A number of different approaches to the problems of planning and scheduling contributed to the development of the subject of these papers. Imperial Chemical Industries and the Central Electricity Authority (now the Central Electricity Generating Board) were using the basic concepts in Britain by 1955 and 1957 respectively. In 1958, two of the most widely known systems of analysis, PERT and CPM, were announced in America. PERT, standing for Program Evaluation and Review Technique, evolved from the need to develop an improved method of planning and introducing the Polaris project. In the military environment the main consideration was time reduction, but a parallel development in the civil field by the Du Pont Chemical Company sought to achieve a balance between cost and time. This research yielded CPM, the Critical Path Method, which produces either the cheapest way of executing the project or alternatively the cheapest way of doing the job as quickly as possible. Not only have PERT and CPM lost their original connotations, but a multitude of other expressions have been developed.

Some areas of application

The object of this section is not to give a comprehensive survey, but merely to indicate to the reader the scope and range of potential application. It is inevitable that a survey as brief as this will omit to mention many firms with wide experience of the techniques, and indeed will fail to indicate the width of experience within any one firm.

A number of civil engineering firms have started to use the techniques only because the firms which invited their tenders insisted on the production of an appropriate network. This state of affairs arose from the pioneering work done in England by leading firms like Richard Costain and John Laing Construction. New manufacturing and chemical process plants have been designed and commissioned using CPA by firms such as British Oxygen Company and Distillers Chemicals and Plastics Ltd. Housing developments by Croudace, and other leading builders, have been controlled through Network Analysis. During 1966 a firm reconstructing a shop in Cologne exhibited its network diagram on the protective hoarding and brought it up to date so that passing members of the public could see the progress being made.

Applications in the other engineering fields include locomotive maintenance by the British Railways Board and plant maintenance at the Steel Company of Wales. Apart from using networks on construction projects, the National Coal Board controls pit operations by the routine use of a planning network at each colliery. Eight-year programmes for all new generating stations are being networked currently by the Central Electricity Generating Board. The United Kingdom Atomic Energy Authority uses Network Analysis for major construction projects and plans reactor overhauls using resource scheduling. British Petroleum uses the technique for refinery maintenance.

The British Coal Utilization Research Association has examined the ways in which the network concept can be utilized in conjunction with the uncertainties inherent in the research and development fields. Alfred Herbert, a leading machine-tool manufacturer in the UK, uses Network Analysis in the design and production of new lines. Unilever uses the technique to plan the development and launching of new products, including packaging, advertising, and consumer research activities.

An interesting application occurs when a network is drawn for the installation of a computer; the analysis of the network is frequently undertaken using another computer. In a similar manner, networks have been drawn to improve a number of clerical procedures, notably in the accountancy field. The British Aircraft Corporation is among the firms using small standard networks which are integrated into a large network for specific projects.

The firms mentioned above are known nationally. However, there are a great number of quite small firms which have utilized the techniques in many of the ways described. The size of the networks and the complexity of their analysis are more dependent on the nature of the project than on the size of firm or even the financial value of the work. Sometimes it might cost more to analyse the project than might be considered appropriate. However, when a task is repetitive, the initial expenditure of effort can repay handsome dividends. One repetitive task which has benefited from Network Analysis is a certain surgical operation for which the time has been reduced significantly.

Some benefits arising from the techniques

What kind of benefits can one expect from the application of Network Analysis? A firm just starting to use it will probably derive a lot of benefit from the need to think about all aspects of the project in advance of its commencement. The manager who does not plan at present, or whose plans frequently seem to go askew, will find no haven in Network Analysis unless he is prepared to be thorough. However, with comparatively little effort he will find this is at least a way in which complex operations can be planned efficiently. His efforts, together with those of his colleagues, will produce an integrated approach to projects, and the responsibility for each part of the work will become clearly established. Information will be provided in a form which allows each manager to control his own tasks by exception rather than in total.

It is frequently thought that Network Analysis is confined to the planning of the construction phase of a project; the preceding section has indicated that it may be used for a variety of purposes including the planning of the planning itself. Many projects may be planned and executed in a variety of ways. What is the 'best' way? There may be a number of conflicting criteria against which each method must be measured. Networks enable these alternative methods to be described and evaluated both quickly and accurately, and then management may select the most appropriate approach to the job.

Network Analysis may also be used as a method of project control by updating the network as the project proceeds and thus anticipating hold-ups in time to take preventive action. Part III describes extensions of the basic technique which provide means of resource and cost control.

A number of other benefits of using these techniques will emerge from the following chapters. We do not claim that Network Analysis is a panacea, but during the past few years many firms have found it provides a most powerful addition to their armoury of management science subjects.

2 · Introduction to the Basic Method

D. WILLIAMS

The steps in network analysis

It is helpful to think of Network Analysis as a number of steps.

    (*i*) Understand the logic of the strategy
    (*ii*) Construct the network
    (*iii*) Obtain activity duration estimates
    (*iv*) Identify the critical path
    (*v*) Satisfy primary objectives
    (*vi*) Satisfy secondary objectives
    (*vii*) Repeat (*i*)–(*vi*) for alternative strategies
    (*viii*) From (*vii*) select the most appropriate strategy
    (*ix*) Install
    (*x*) Maintain

The first eight of these steps are considered in this section. Installation and maintenance are considered in Part II.

Only when one has attempted to represent the logic of a situation by a network does one realize the limitations of one's understanding of that logic. Therefore it is essential to obtain a clear picture of the logic in order that the network may be drawn correctly. At this stage no information is required about the durations of any activities.

Activities and events

Most tasks consist of a number of interrelated activities. Each activity may be represented by an arrow. The length and slope of the arrow are independent of the duration of the activity. The function of the arrow is to represent the logical relationships which exist between the particular activity and the other component activities of the project. It is found useful to present the completed versions of a diagram with the arrows flowing from left to right.

7

The points at which arrows converge or diverge are called events or nodes and are represented by circles or ellipses. An event consumes neither time nor any other resource. Each event is given a reference number and misunderstandings are reduced if the event preceding an activity always has a lower number than the event succeeding that activity.

Consider a statement concerning three activities which occur in the middle of a project: Job N follows Jobs L and M. *Figure 1* is incorrect since it implies that Job M follows Job L. It is always difficult to ensure that the given logic, and only the given logic, is represented on the network. *Figure 2* is one way of representing the logic correctly.

Figure 1   *Incorrect logic*

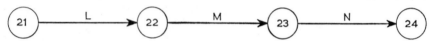

Figure 2   *Correct logic*

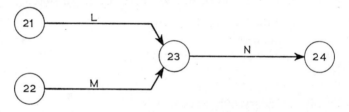

It is useful to draw each activity with a horizontal component along which the activity description may be written. It must be emphasized that it would not be correct to deduce from the network either that L and M were of equal duration or that they are necessarily scheduled to start or finish at the same time. The question of scheduling must not be considered until the network has been completed.

Dummy activities

The very simplicity of the network diagrams has led many people into erroneous representation of the logic of a project. Consider the following statements:

> Jobs D and E may proceed in parallel
> Job D must be completed before either F or G may commence
> Job E must be completed before G may commence

Figure 3 is incorrect because it additionally shows that Job F cannot commence until Job E has been completed. In order to represent only the given conditions it is necessary to introduce the concept of a dummy activity. This is defined as a logical link which represents no specific operation and is represented by the broken arrow shown in *Figure 4.*

Figure 3   *Incorrect logic*

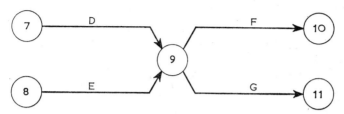

Figure 4   *Correct logic using dummy activity*

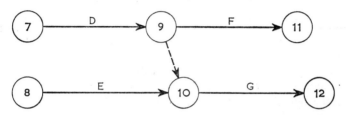

A dummy may also be used to distinguish between two activities such as B and C shown in *Figure 5*, neither of which may start until A has been completed, but both of which must be completed before any subsequent activity may commence. In an activity listing, both B and C would appear as activity 2–3. *Figure 6* shows one way in which a dummy may be used for this purpose.

Figure 5   *Ambiguous representation*

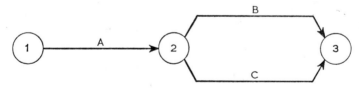

Figure 6   *Clear representation*

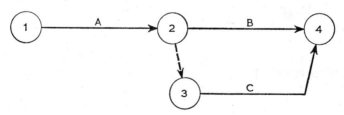

Assumptions of strategic logic

So far we have discussed the means of representing a given strategy. There may be a number of alternative strategies, the one assumed should always be specified and this specification should accompany the network. An example using a situation which frequently causes difficulty may underline the point. Consider the way in which one could represent the provision of a piped water supply to a village. We are concerned only with the trench-digging, pipe-laying, and backfilling. One way of representing this work would be as in *Figure 7*. This strategy shows that all of the trench should be dug before any pipes are laid and they must all be in position in advance of any backfilling.

Figure 7   *First strategy for pipe-laying*

In many sequential jobs of this nature, the first strategy is unrealistic. A more common strategy is shown in *Figure 8* where the length of trench has been divided into three sections, a, b, and c. It is given that three separate gangs will be available, one responsible for digging, one for laying, and one for backfilling. We can assign various bands of the network to the gang which is responsible for each facet of the work.

Figure 8   *Second strategy for pipe-laying*

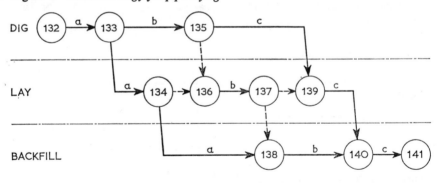

This particular strategy also assumes that any one operation will be completed in the first section before the same operation is started in the second section, and so on. It is obvious that in any one section the trench must be dug before the pipes are laid and that there is no point in backfilling until after the pipes have been laid. Clearly the number of sections could be increased under an alternative strategy.

The methods of splitting jobs up into sections is sometimes summarized by a technique called 'laddering'. This technique uses one arrow for each of the types of work, corresponding to the bands of the network in *Figure 8*, but joins them with 'lead' and 'lag' lines to indicate that the work proceeds mostly in parallel (see *Figure 9*). The 'lead' lines indicate the time which must elapse before the subsequent activity can start and the 'lag' lines indicate the time after the end of the first activity that the second will finish, etc.

It is frequently possible to question the apparent need for an activity to be completed before another one may start. The reversal of such an assumption will frequently yield a diagram such as *Figure 8* or *9*.

In other cases there may appear to be a choice about the order in which

Figure 9 *Laddering*

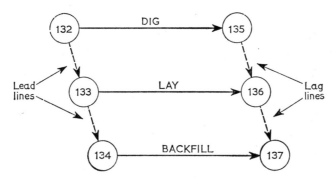

component jobs may be done. An example concerns the battery which operates auxiliary equipment in a locomotive; the battery may be topped up before or after the associated circuits are tested. There are two ways of approaching such problems. One may either specify the order in which the jobs are to be done, or one can say that one unit of a resource, 'battery', is required for each job and that there is a maximum availability of one unit of 'battery'. The latter method can only be adopted if one of the resource allocation techniques is being used (see Chapter 7).

Activity durations

The construction of the network required no knowledge of activity durations; but in order to carry out the basic analysis of the network it becomes necessary to obtain estimates for the duration of each activity. The durations may be written on the arrows.

If the activities have been subjected to Work Study, objective times will be available. In other cases the nature of the work is understood fully and realistic estimates can be provided from previous experience. Occasionally there may be difficulties in making time estimates; this problem is considered in Chapter 8.

Experience confirms that a certain latitude is possible in the initial activity estimates without prejudicing the ultimate project duration. The reason for this lies in the facility which is provided for a regular review of the work while it is in progress. Not only may more recent estimates be incorporated at that stage, but also the effects of management controls may be evaluated. These questions are discussed further in Part II.

Event times

Events consume no time, they merely mark the transition from the state when some activities have terminated to that when other activities may commence. There may be some flexibility about when this transition may occur; therefore it is necessary to calculate the earliest and latest times at which the event can occur. The flexibility is known as event slack and is defined as the latest event time minus the earliest event time. Normally when the slack is zero the event is critical and cannot be delayed without the project being delayed. The exception to this rule occurs when time restraints are imposed (see below, pp. 15–16) creating some events with negative slacks. Such events may be described as hypercritical.

When the event has only one preceding activity its earliest time is defined as the earliest time of the preceding event plus the duration of the intervening activity. Where there are several preceding activities it must be remembered that all must be completed before any subsequent activity may commence. Thus the alternative earliest times based on each of the preceding events must be calculated and the largest one selected. Starting with time zero for the earliest time of event 1, *Figure 10* shows the earliest times at the events located in the first position within the rectangular boxes adjacent to the events.

Figure 10   *Earliest event times*

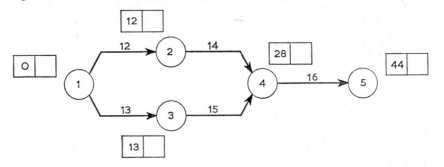

The latest time is defined as the latest time of the succeeding event minus the duration of the intervening activity. When there are several succeeding events the arithmetic is carried out for each of them and the smallest one is selected. At the last event the latest time is taken to be the same as the earliest time. *Figure 11* shows the latest times superimposed on *Figure 10*. The critical events are 1, 3, 4, and 5.

External considerations may impose time restraints which may modify either the earliest time, or the latest time, or both. Such restraints are discussed below.

Figure 11 *Latest event times*

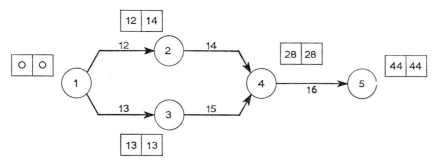

Activity float

Not only are the event times of interest in themselves, they also enable one to calculate the two times associated with the activities which are defined below.

Earliest Start Time (EST) = Earliest time of its preceding event.
Latest Finish Time (LFT) = Latest time of its succeeding event.

These two times establish the limits between which the activity may be scheduled. Two further times can now be derived.

Earliest Finish Time (EFT) = EST + duration of activity.
Latest Start Time (LST) = LFT − duration of activity.

The flexibility available for the scheduling of the activity can be measured by one of several floats which are defined below.

$$\text{Total Float} = (\text{LFT} - \text{EST}) - \text{duration}$$
$$\text{or LFT} - \text{EFT}$$
$$\text{or LST} - \text{EST}$$
Interfering Float = succeeding event slack.
Early Free Float = total float − succeeding event slack
Late Free Float = total float − preceding event slack
Independent Float = total float − (preceding + succeeding event slacks)

Total float is the one most frequently used in network analysis. *Figure 12* illustrates total float with the activity scheduled at three different times.

When the total float is zero, the activity is critical and any delay in the activity would normally delay the completion of the project. The exception to this rule occurs when time restraints (see the next section) are imposed, creating negative total float. The activity is then described as hypercritical. The critical path is a path from the start event to the end event, the total

Figure 12 *Total float*

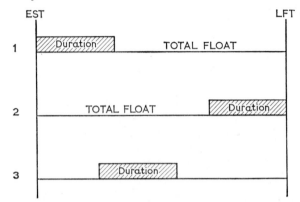

duration of which is not less than that of any other path between the same two events. The critical path is shown by the double arrows in *Figure 13*.

Figure 13 *Critical path*

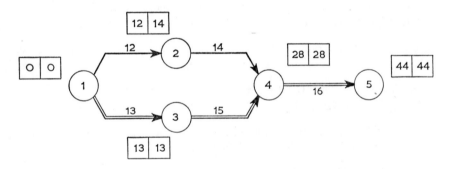

Working-day calculations

The calculation of event times and floats has been described in terms of absolute time; a slight difficulty arises when we come to assign real dates or times to a network. Consider the example of *Figure 10* supposing that the durations are in working days and that work is to be carried out on a seven-day-week basis. The calculation for activity 1–2 based on an absolute starting time 0 gives an EFT of 12. If the day on which work begins is called Day 1, then work will indeed finish on Day 12. The EST for the next activity (2–4) is calculated as 12 (from the definition on p. 13); however, it is apparent that work cannot actually start until Day 13.

In practice, the difficulty is not serious; however, some people prefer calculations to refer to working days, which means adding 1 time unit to all starting times. Many computer programs can output activity lists using a working-day numbering system with corrected starting dates. Allowance must also be made for the length of the working week if calendar dates are to be used (see p. 28).

Types of restraint

Factors external to the problem represented by the network may need to be superimposed on the analysis. If these occur through conditions in a related network the approach given below for a single level network can be extended to the multi-level networks discussed in Part II.

In single-level networks it may be necessary to fix target dates for the latest completion of tasks, or for the earliest start of tasks, or to make allowance for the fixed internal meeting of the board or one of its committees. Broadly speaking, restraints will be such that the range of dates for an event is strictly limited or that two events must be separated by not less than a given period.

One may deal formally with restraints by including them in the network logic and assigning them algebraic durations. The network is first analysed ignoring the restraints and then their impact is assessed. Two examples will clarify this point.

Figure 14 *Algebraic activity duration*

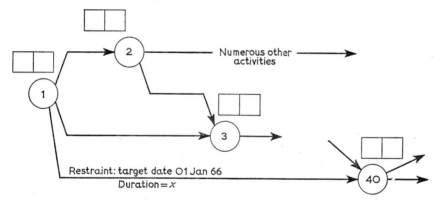

First consider the imposition of a target date on event 40 in *Figure 14* of 1 January 1966. It is not possible to specify the relationship between the calculated event times and this target date until the start time for the network is related to the calendar. Represent this restraint as activity 1–40 and assign it duration x. Once the times for event 40 have been calculated they can be compared with x. Four possibilities exist:

(*i*) x greater than latest time event 40.
   No problem.
(*ii*) x less than latest time event 40 but greater than earliest time event 40.
   Slack reduced, but no problem.

(*iii*) x equal to earliest time event 40.
     Event is critical.
(*iv*) x less than earliest time event 40.
     Event becomes hypercritical.

Management's attention should be drawn to cases (*iii*) and (*iv*). An investigation should be made which challenges the need for the specified target date before anyone rushes off to organize remedial action.

The second example is concerned with the need to separate two events by some interval. This interval is either known exactly or can be expressed in a simple formula. The period between the sitting of a middle management committee and the endorsement of its minutes by a senior executive is usually known and of fixed duration and so no new problem is introduced.

When, however, an item is submitted to the middle management committee for inclusion in its agenda, there may be no guarantee that the item will be reached at the next four-weekly meeting. Let p weeks be the time from commencement of document preparation to the committee meeting date next after document completion. Let the item be reached at the nth meeting after submission; if, once reached, the item is completely dealt with at the same meeting, the duration for the activity covering preparation, submission, and approval, could be expressed as $p + 4(n - 1)$ weeks.

The value for n may not be known when the network is first drawn, but as the relative importance of the items competing for the agenda becomes clearer it will be possible to provide realistic figures for n.

Management objectives

Management objectives will vary from project to project and the particular variant of network analysis should be chosen to suit these objectives. If time minimization is all important, straightforward network analysis may be sufficient. The uses of cost minimization and control are discussed in Chapter 10 and resource allocation in Chapter 7. If the project has a substantial research or development content the methods of Chapters 8 or 9 may be relevant. It may be important to take more than one of these aspects into account, in which case secondary objectives must be introduced, possibly in the form of restraints.

Where there are several possible strategies for carrying out a job it may be very important to make the correct choice of strategy to achieve the desired objectives. A network can be drawn for each strategy and the success with which it meets the objectives can be evaluated. Thus a factual basis for comparing strategies is available.

3 · The Method of Potentials

J. A. CARRUTHERS

Introduction

The 'Method of Potentials' is another approach to project planning and control. It was developed in France from studies in Scheduling and Graph Theory by Monsieur Roy, about the same period as CPM and PERT, but quite independently of them. Superficially different from the conventional network approach, the method nevertheless is basically the same. Critical path methods consider activities as arrows, using events as the instants which start and finish them. On the other hand, the Method of Potentials considers events (or the starts of activities), using the actual activities or arrows as links between events. Thus the use of nodes and links to denote events and activities respectively in CPA is reversed in the Method of Potentials.

Network relationships

A conventional critical path network represents a project as a group of activities. The start of each activity is an event, which is logically linked to the completion of its logically preceding activities. Let the start of the jth activity be at a time t_j; then for all logically preceding activities, i;

$$t_j \geqslant t_i + d_i \tag{1}$$

where d_i is the duration of the ith activity; that is, t_j has lower bounds. The earliest time at which t_j can occur is obviously the maximum of the lower bounds set by the preceding activities, i; thus:

$$\min t_j = \max{}_i (t_i + d_i) \tag{2}$$

From the above relationships alone, one can carry out a complete time analysis for a project. Consider the activities in *Figure 15*, and their durations, necessary for the construction of a small building.

17

Figure 15 *Activities to construct a small building*

Activity label	Description	Duration
a	Clear and set out	4
b	Lay foundations	6
c	Lay drains and main services	4
d	Build shell	9
e	Build roof	4
f	Plumbing	4
g	Carpentry	4
h	Electrical	3
i	Glazing	2
j	Painting	4
k	Garden	4
l	Snagging and clearing out	3
m	Plastering	4

To carry out the time analysis, the starting time of each activity is tabulated, and, in a column below it, is written each logically preceding activity together with the time which must have elapsed from its start until the current activity can start. *Figure 16* shows the table for the above example

Figure 16 *Tabular representation of start events*

a	b	c	d	e	f	g	h	i	j	k	l	m	F
0 D0	a4	a4	b6	d9	e4 c4	e4	e4	e4	m4	c4	k4 j4	f4 g4 h3 i2	13

Two events additional to *Figure 15* have been inserted within the body of the table, one called D (*début*) for the start of the project, and the other called F (*fin* or finish) for the end of the project entered as an additional final column. No further diagrams are required to complete the analysis. Nevertheless it may be helpful to the reader to show *Figure 16* in network form, see *Figure 17*.

Each activity, or strictly its start event, is in a labelled 'box', instead of being an arrow in a conventional network diagram. The links between the boxes are the logical network relationships, and the numbers entered on the links represent the minimum times which must pass between the events. To continue the time analysis from *Figure 16*, the start event, **D**, by convention is known to start at 0, and hence the number 0 is entered to the left of **D** wherever it occurs in the table (under column a, in *Figure 16*). Hence it is seen that event a will occur at time 0 + 0, i.e. 0, and this 0 is entered at the

Figure 17　*Diagram of potential relations for a project*

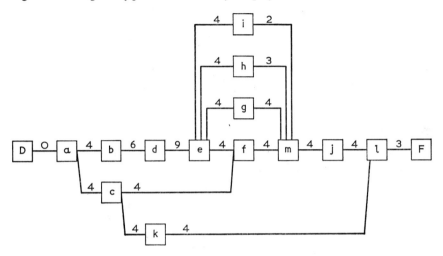

head of column a (see *Figure 18*); the figure 0 is then entered to the left of every other entry of a in the table (i.e. under columns b and c). Proceeding thus, as soon as all the entries in the left hand of each column have been filled, the starting time can be determined; thus, the starting time of f is the maximum of 19 + 4 (from e) and 4 + 4 (from c), i.e. 23. *Figure 18* illustrates *Figure 16* when the time analysis has just finished column f.

Figure 18　*Intermediate stage of time analysis*

a	b	c	d	e	f	g	h	i	j	k	l	m	F
0	4	4	10	19	23								
0 D 0	0 a 4	0 a 4	4 b 6	10 d 9	19 e 4 4 c 4	19 e 4	19 e 4	19 e 4	19 e 4	m 4	4 c 4	k 4 j 4	23 f 4 g 4 h 3 j 2	1 3

It is a simple matter to complete the time analysis, starting from either end of the project, and to determine slack and similar information, as in conventional critical path analysis. Thus *Figure 16* represents all the logical and time dependencies in a project as completely as does a network diagram.

Comparison with critical path methods

The essential difference between the Method of Potentials and conventional critical path methods is that on paper, the events and their tabulation are central to the Method of Potentials while the activities and their arrow diagram representation are central to critical path methods. In computer terms, this difference effectively disappears, since both methods require an event list on computation. In fact, some current programs can input data listed by either technique, and can output results either by the same technique or by its alternatives. In human terms, this difference effectively disappears again, since both methods require an activity list as an action list for project control. Some advantages and disadvantages of the Method of Potentials over conventional CPA in constructing and analysing a project network, are listed below.

Advantages

(*i*) No network diagram is strictly necessary, since all the logical relationships are present in the table of events. (Nevertheless, a diagram may still be drawn as an aide-memoire, especially in practical networks of more than 100 activities.)

(*ii*) No 'dummies' are necessary (since all activities are unique nodes of the network, and all logical links are contained in the event table).

(*iii*) Alteration of project logic is simple, requiring only addition or subtraction of items in the event tables, or links in a network diagram. This may be especially useful when project network is inherently uncertain or in the earliest planning of a project.

(*iv*) Certain situations, such as several activities which may proceed simultaneously between the same start and end events, are far more easily represented, in view of (*ii*) above.

(*v*) The symbolism of activities at nodes is consistent with established Work Study notation.

Disadvantages

(*i*) It lacks the visual appeal of a network diagram, and thus a diagram is often still stipulated.

(*ii*) Certain situations, frequently encountered, where some important event has several activities entering it directly and several leaving it directly, are much clumsier to represent than in conventional critical path analysis. In such situations the insertion of a 'dummy' activity between m initial and n logically succeeding activities may be useful in reducing the number of diagramming links from $(m \times n)$ to $(m + n)$.

Additional features

The Method of Potentials allows relations of different forms to the normal network relations, such as:

$$t_j \leqslant t_i + d_i \qquad (3)$$

This type of relationship makes possible the inclusion of restrictions of the type: concrete must be poured within a given time of mixing. Combinations of (1) and (3) furthermore allow the possibility of stipulating that activities shall be simultaneous.

Part II

Early Problems of Implementation

An organization which makes a policy decision to adopt CPA as a planning technique immediately faces a number of problems. The first problem is what sort of personnel are needed and what project should be selected for the first application of CPA. This is an important decision, for a poorly selected, unsuitable project can completely discredit CPA as a planning tool. Other, more technical, problems that often arise in the first applications of CPA are how much detail to include in the network, whether to analyse the network by hand or on a computer, and how to control the project satisfactorily once it is in progress.

The papers in this Part consider some of the problems of implementation and their possible solutions. These solutions are offered only as a guide; individual companies or organizations will find they have to modify the suggested methods to meet the needs of their particular projects and organizational structure.

The problem of whether simple CPA is sufficient, or whether resources, costs, or probabilistic times should be included, is deferred to Part III. It is strongly advocated that several time-only analyses should be successfully carried out before any more complex analyses are attempted.

4 · Introducing CPA into an Organization

R. COKER

Personnel and education

The first need of a company which has decided to use network planning methods is for trained staff. This can be fulfilled in at least four ways:

(*i*) Employ consultants to introduce and implement network methods.

(*ii*) Advertise for an experienced practitioner.

(*iii*) Select someone from within the organization and send him to a training course.

(*iv*) Select someone from within the organization who is capable of teaching himself the rudiments of network planning.

These methods are in order of decreasing initial cost. But before selecting (*iv*) it must be borne in mind that, no matter what calibre of man is given the task, a period of at least one year will be needed before any network method begins to show dividends. Method (*iii*) is not much better than (*iv*), only more expensive. Method (*i*) tends to be very expensive and will still require either selection of personnel from within the company or employment of additional, probably experienced, personnel. In conclusion, method (*ii*) probably gives best value for money, but method (*i*) is quicker.

Assuming now that the company has employed a man with CPA experience allied to some appropriate industrial experience – what then? It is fair to say that unless CPA is understood in principle throughout an organization it will fail to be effective. This is because CPA, as a planning tool, is not confined to the planner's office, but involves everybody who is in any way linked to the project being planned. Any attempt to use CPA must, therefore, be preceded by a series of appreciation talks, firstly to top management, and then successively to the lower echelons until all managers likely to be involved in its use have been given some education in the principles of CPA. These appreciation courses need be of only one or two hours' duration, but more time

25

should be spent (up to a whole day) with those managers who will eventually be expected to control projects by CPA.

The first application

The next step is to select a project to plan. It may be suggested that a network of a job already completed would immediately show the value of CPA, but, in general, personnel can find no enthusiasm for such an historical exercise. A good initial project should be sufficiently complex to make a network more than just a series of connected arrows. On the other hand, the project should not comprise more than 100 or 200 activities, so that the network may be portrayed on a single sheet of paper.

Experience has shown that there are a number of pitfalls into which the less experienced practitioner may fall. The strict logic of some projects often indicates that, say, a dozen activities could start simultaneously, but in practice these activities may well be done by one man. Since he cannot work on all activities simultaneously the project cannot be done in the time indicated. In other words the project is 'resource limited'. A project of this nature is not recommended as a first project because, if the limitation is not realized, the project will go over the target date despite all remedial efforts and thus CPA may fall into disrepute. If, on the other hand, the resource limitation is recognized, the network can be redrawn to incorporate the limitation, or one of the resource allocation methods may be used. Neither of these alternatives is recommended for those about to start their first CPA application.

Another problem is the question of policy in networks. It is quite often possible to state in network form 'this is how we would like it to happen', but the alternative 'this is the strict logic of the situation', can also be drawn. So long as it is clearly understood that part of the network is dictated by policy and not logic it is a reasonable approach, but again it is not recommended that this type of complexity should be built into the first network.

Once a project has been selected by the chief planner, the agreement of the project manager must be obtained. It is not wise to force CPA on a manager. The next step is to hold a meeting with all those who will be both contributing to the formation of the network and estimating times for activities. The purpose of this meeting is to acquaint them with the decision to use CPA on this project, to outline what will be happening during the next few weeks, and to explain what results will be forthcoming and how they will be used. All concerned should be encouraged to ask questions freely so that everyone understands what is to be done.

The next step is to draw the network. There are two possible methods:

(*i*) Get all concerned to sit round a table and start drawing.
(*ii*) Let the expert draft out a complete network in co-operation with the project manager, and subsequently discuss it in detail with each person who is responsible for an activity or group of activities.

Method (*i*) is the better for the first network, but for subsequent networks, or even subsequent drafts of the first network, method (*ii*) is better, and saves time. As the network progresses emphasis must be continually placed on the accuracy of the logic embodied in the diagram. As each event is reached the question must be asked: does this activity *really* depend on this, this, and this?

It is often impossible to depict the rigid logic of a project without going into fine detail. The internally-trained planner may be found lacking here, since experience of many networks is needed before a planner is able to find the 'right' level of detail for any particular job. It is often better to overrule the rigid logic in order to produce a 'workable' network. This problem is considered further in Chapter 5.

During the network-drawing phase, people contributing to the logic will undoubtedly comment on the fact that the planner has not considered how many men will be available. Neither has he considered how the resource requirements of the activities may interfere with one another. In answer to this, it must be constantly emphasized that the network approach is to discover the logical sequence of activities first and then, by putting in time estimates and calculating, to discover the juxtaposition of the individual activities. Then, and only then, are questions of resource limits, timing, and interference of one activity with another dealt with in the light of the discovered critical path and the associated non-critical activities.

When the network is thought to be complete, the duration of all activities must be estimated. This may be a relatively easy task, especially in construction projects where, for instance, the rate of pouring concrete is well established. In other projects, particularly those with a research or development content, it may be extremely difficult. When estimating is difficult, it should be emphasized that any *reasonable* estimate is sufficient for a first trial, since it enables the critical path through the network to be calculated and thus throws the attention of management onto the appropriate activities. The estimates can then be reconsidered as more information becomes available.

There is often a feeling that, once an estimate has been made, failure to meet that target will be recorded as a black mark against the individual responsible for the completion of the activity. To offset this feeling it should be emphasized that out of all estimates made only a few will be achieved in reality, the remainder being either over- or under-estimates. However, there may still be a tendency for the estimator to add 10 per cent, but if this estimate comes from a section head, who got it from a foreman, who spoke to the charge hand, and if they all added 10 per cent, the net result is unrealistic. One method of overcoming this, which is considered further in Chapter 8, is the use of three time estimates. An alternative approach is to break down the critical activities, discovered in a first rough calculation, into more detail to arrive at a more reliable project duration.

In the process of estimating, changes in logic are quite often discovered. The task of putting a time to any activity forces the estimator to consider the components of the activity, and thus may bring to light a more accurate description of the logic associated with that activity.

When the network is in its final form and all activity times have been estimated, the event times must be calculated. It is recommended that this should be done manually on the first project, if only to permit people to see how simple it is. Manual calculation does, of course, necessitate working in time units (days, weeks, months, etc.). This means that a calendar must be set up to translate, say, day No. 123 into Wednesday, 25 May 1976. This complication, while tiresome if a great number of projects are being controlled, is of little consequence in a first project which has been selected to have only one or two hundred activities.

The next step in this first application is to interpret the results. First the project manager should recognize the critical path. In a very small network this could be sufficient to ensure that the project is completed on time, but where many details have to be constantly reviewed it is recommended that all activities should be listed in order of their start dates together with their associated float. In this way the manager can keep track of every activity as it is due to start. Although the process of sorting and collating the data can be achieved much more quickly by computer, it is not recommended that a computer should be used on an organization's first application of network analysis.

Project control

For the first project it is suggested that every activity should start at its earliest starting time and that float should be dispensed only with caution. As experience is accumulated more sophisticated uses of float can be tried, for example proportionate subdivision of total float to all activities in a chain. The question of who should know what float is available largely depends upon the management structure. It is reasonable to say that the personnel actually doing the task should not be told what float exists. Their administrators, or managers, should know more and more detail regarding float the nearer they are to the project manager.

Controlling the project consists of two main tasks: issuing instructions, and taking action if instructions cannot be complied with. Depending on the type of project being controlled, instructions will vary from a bar chart showing next week's work to a page or section of the computer print-out. Instructions for the first project (which would probably not be processed by computer) might be circulated as a typed list of the earliest starting dates of those activities near the start of the project, with further lists as the project unfolds. All instructions issued by the project manager must be accompanied by the reminder that the target dates shown are estimates, and if it becomes obvious at any time that they cannot be met, the individual responsible must immediately notify the project manager so that remedial action can be taken if necessary.

As a project proceeds the progress of individual activities should be marked regularly on a network. It will thus be obvious when an activity has taken longer than estimated, and the effect of this overshoot on the whole project will be easily discovered if the network is small. Control of progress on a large project may be more easily achieved by regular computer updating runs. Updating runs and other aspects of computer use are dealt with in Chapter 6.

Senior management involvement

During the life of most projects a number of decisions need to be made by the senior management of the organization. Once the relevant facts have been assembled, the actual decision process takes very little time and so is represented merely as an event on a network. However, in most organizations it is not possible to get top management to take decisions as quickly as that; there is often a full in-tray to be gone through, or a delay while the executive thinks it over. If the decision event has slack, it is reasonable for some float to be absorbed. On the other hand, if the decision is critical, top management must be made aware of this, and of the implied result of any delay. The obligation to make a quick decision will then be apparent, as will the consequence of delay.

Subsequent projects

As experience accumulates it should become feasible to tackle larger and more complex projects. Large projects however may well involve outside consultants, contractors, and suppliers. In their own interests the company initiating and controlling the project should attempt to ensure that all concerned have sufficient education in the technique. The special problems of large projects are considered in Part IV.

ACKNOWLEDGEMENT

The author is indebted to Mr E. L. Buesnel of Unilever Ltd for his invaluable comments on the first draft of this paper.

5 · *Optimum Level of Detail*

H. WALTON

Types of network

The optimum level of detail for a network derives naturally from the use to which the network is to be put. Networks are drawn for use in two main areas of analysis and control:

(*i*) Plan-only projects in which the network is used to assist in the overall development of ideas.
(*ii*) Projects in which the network is used to plan and control a project from beginning to end.

The plan-only type of approach is often concerned more with ideas than with actual work. As a result of this the level of breakdown of activities bears a relationship to the individual ideas developed. A break in concept or in responsibility for an area of endeavour will indicate the need to change from one activity to the next.

Plan-only networks are of use in the ill-defined phases of the development of new products, or in any area where the correct correlation of ideas and actions is of more importance than a detailed study of timing and progress. Some areas to which the plan-only network can be applied are: administrative reorganization; the study, in conjunction with the flow process chart, of the technique of starting up and shutting down process plant; the preparation for important negotiations in which it is necessary to foresee the likely trends of an argument and to prepare appropriate counter-arguments; and domestic plans, either relating directly to the family or in the development of ideas for plays, films, etc.

In all other areas of application, the network is used to depict a finite project in which the individual contribution of many departments, sub-contractors, etc., must be correctly correlated. Under these conditions, in all

31

except the very largest projects, one becomes aware of three planning levels, namely:

Level   I Definable areas of operation which may be geographical, technical, or other. These areas are labelled to assist comprehension of the total work content.

Level II Subdivisions of each Level I section into the functional or technical activities necessary to complete the work in that section.

Level III Subdivision of certain single activities from Level II into local subnetworks.

The activities depicted at Levels II and III divide naturally into 'quanta of responsibility'. These quanta of responsibility are individual areas of work depicting one sphere of responsibility or one sphere of skill. The magnitude and depth of detail of these areas depend on their relationship to the project manager. The manager will normally wish to exert personal influence over the section depicted by the activity, but not to exercise control within the quantum of responsibility.

Monitoring of progress

The depth of detail in the network will automatically emerge from a realization of these individual areas of responsibility, each event or nodal point signifying a change in responsibility. This same realization automatically yields the ability to monitor progress by obtaining reports from the individuals responsible for each of the activities.

If such monitoring is not possible the network has been incorrectly drawn, and locally, at least, the depth of detail is inconsistent with the individual responsibilities. The ability to monitor the activities in a network is a prime indicator of the depth of detail required. Here and there it may well be necessary to derive a Level III subsidiary network for local control in order to yield proper reports for a single Level II activity.

Factors affecting the level of detail

While the concept of a quantum of responsibility is the main indicator of the optimum level of detail required in a network, there are nevertheless other factors which give rise to subdivisions of activities within this major definition.

The logic may call for a second activity to start part way through an initial activity, thus necessitating a split in the arrow depicting the initial activity, i.e. a nodal point must be inserted to cater for the start of the second quantum of responsibility while the first is still progressing. Even though the quantum of responsibility may remain uninterrupted, there may be a change in resource requirement part way through it. This gives rise to an intermediate node to satisfy the needs of resource allocation. This only applies, of course, when resource allocation or smoothing is being undertaken. An activity within a large project may be on or near the critical path; as a result of this, it may be thought prudent to introduce what is virtually Level III detail by breaking up the activity at that point into finer quanta.

In order to counteract the tendency to overcomplicate networks with too much detail, activities should, in general, be kept as comprehensive as possible and, therefore, as few as possible. Excessively long or costly activities can result from this approach, however, and the need to control these and report progress on them will call for a greater depth of detail at such points.

In general, as much forward information as possible should be included in the network, even extending several years ahead. This will lead to a more reliable total evaluation. As long-term projects proceed, however, there will inevitably be a build-up of knowledge for the immediate future as the project progresses. The closer one approaches a particular area of work the more clearly defined become the quanta of responsibility.

Depending on the ultimate use to which a network is to be put, it may be permissible to employ a looser or tighter logic to depict it. If full monitoring and control is to be effective, then tight or rigid logic is essential in order to yield clarity in reporting. Loose networks are justified when rapidity of preparation is essential and when they are merely to be used to depict the content of a project rather than as a basis of detailed control. Such networks may be drafted at the tendering stage and may even be used during the early days of a main project until the full detailed network is complete.

No attempt should be made to minimize the number of activities in a network by leaving out those which are considered to be unimportant and to have considerable float. These can easily be forgotten and their omission prevents complete evaluation of the project.

Laddering techniques as a means of simplifying networks are deprecated since, in the final result, they do not in fact lead to simplification nor do they give rise to any significant difference in the depth of detail.

Occasionally, considerations external to network analysis may have a determining effect upon the level of detail. Increased accuracy of timing may be required, thus calling perhaps for greater detail. Alternatively, the sheer magnitude of the task of monitoring a network may call for a decrease in the amount of detail.

Conclusion

While it is difficult to be precise as to the depth of detail required for plan-only networks, the depth of detail required for project-controlled networks emerges naturally from a consideration, first, of the need to portray the logic sequence of the network, and, second, of the need for the project manager to be able to assess overall progress by reference to each unit of responsibility.

ACKNOWLEDGEMENT
Many of the ideas put forward in this section stem from a report, dated 16 September 1964, prepared by the Study Group's Working Party on Optimum Network Size and written by J. N. D. Scott.

6 · The Use and Abuse of Computers

D. WILLIAMS

Introduction

Much has been written which might suggest that network analysis is impossible without a computer. The preceding sections have indicated that worthwhile analyses can be undertaken using nothing more than pencil and paper. What therefore is the role of the computer? How can a network be analysed by a computer and what factors should govern the choice between computer and manual analysis?

The program

There are many programs now available and probably only the more sophisticated users of network analysis will fail to find one to satisfy their needs. Under such circumstances a special program can be written. The more normal problem is the selection of the most appropriate program. For simple, time-only analysis there is little difference between programs; the complex programs differ more widely and care must be taken to select the one most suited to the needs of the project. Appendix 3 lists the programs currently available in the United Kingdom with their special attributes, and this can be used as a basis for selection. The two basic methods of resource allocation used by most programs are considered in Chapter 7.

When a program has been selected a form must be completed which provides the computer with information necessary for the program, and specifies what outputs are required. This information varies with the analysis but will include the time units of the durations and, if these are working days, the number of working days per week, a calendar starting date, etc.

Network analysis input

The calculation of the various times and floats in a network requires only a knowledge of the logical structure of the network and the durations of the

Figure 19 *Typical layout of a computer input form*

1 2 3 4	5 6 7 8	9 10 11 12	13 14 15 16 17 18 19 20	58 59 60	61 62 63	64 65 66 67	68	69 70 71 72 73 74 75 76 77 78 79 80					
Start event	End event	Time estimate	Activity description		Dept code	Target time	*	Resources					
								Ref.	No.	Ref.	No.	Ref.	No.
1	8	50	S E L E C T E D		C Ø N			K	1				
1	9	1	A P P R Ø V E		B R D			A	1	K	1		
9	3 5	1 3 0	D E L I V E R		M F R								
1 5	3 5	1 0	T R A I N C Ø		T R G			Z	3				
3 3	5 5	2 0	S P E C I F Y		C Ø N	2 9 5	E	K	1				
4 7	5 7	6 5	D R A W U P		C L I			X	1	Y	1		
1 0 1	1 2 7	6 0	Ø R D E R A /		C L I			W	1				
1 0 0 0	1 0 0 1	5	T E S T T R A		M F R	6 1	S	R	1	S	1	K	1
1 0 0 0	1 0 2 1	5	S E E K U S A		C L I			B	1				

Key * S for Target Start
E for Target End

activities. A list of activities with their start and end event numbers completely identifies the interrelationships between the activities, and their corresponding durations can also be listed. Only this information need be given to the computer as data or input, but frequently it is useful to provide an alphabetical description for each activity. Further information can be given which relates to departmental responsibility, resources, costs, and so on. The way in which such information is used is described in Part III.

Special forms are available for writing out the activity lists; a typical layout is shown in *Figure 19*. Details will vary with the make of computer and the type of installation. Close liaison with the bureau is advised. One line is used for each activity. The form shown in *Figure 19* enables both target times and resource requirements to be entered for each activity as appropriate.

Network analysis output

The basic output of a time-only analysis is an activity list showing the float available for each activity. One of the useful facilities of a computer is its ability to sort out the results into any required order. *Figure 20* shows the activities listed in end event order within start event order. It can be seen that some activities are hypercritical. A sort into total float order would produce a list of activities showing the most hypercritical, those less hypercritical, all with zero float, and finally the remaining activities in ascending positive float order.

There are a number of other sorts which may prove useful. When several departments are involved in a project it is possible to arrange for departmental results to be provided on separate sheets of paper and sorted within each department by float and so on. Resource allocation and cost control programs will have additional outputs corresponding to the analyses described in Part III. By various sorts on the computer and printing of selected information only, it is possible to produce reports at suitable levels of detail for different levels of management. Similarly float, cost, or resource information can be released selectively.

In addition to the purely numerical outputs outlined above, it is possible to produce results graphically and in bar chart or histogram forms if requested. These outputs are particularly useful for the resource and cost analyses described in Part III.

Updating by computer

When it becomes necessary to update a network using a computer the changes are input together with the original data. The program is designed to accept the new information and to discard that which it replaces. Four kinds of changes are usual: those affecting the network logic, resources required, target dates, and activity durations.

Changes in network logic arise when any interrelationships have become

Figure 20 *Typical print-out in I, J order without resource scheduling**

I	J	D	T	*	Description	EST	LST	EFT	LFT	TF	EFF
47	57	65			DRAW UP SITE PLANS	137	294	201	358	157	0
47	94	0			DUMMY	137	201	137	201	64	0
53	109	1	245	E	TEST TYPESET PROGRAM	196	245	196	245	49	23
55	67	165			COMPUTER DELIVERY	152	309	316	473	157	70
57	63	50			APPROVE SITE PLANS	202	359	251	408	157	0
57	97	30			APPLICATION	202	359	231	388	157	106
59	60	0			DUMMY	154	56	154	56	-98	0
60	61	5			COURSE—BOMPS	154	56	158	60	-98	0
61	71	40			PRELIMINARIES	159	61	198	100	-98	0
63	67	65			PREPARE SITE	252	409	316	473	157	0

Key I: Start Event Number   T: Target Time
J: End Event Number   * S for Target Start
D: Duration   E for Target End

* This print-out uses the working-day convention described on pp. 14-15 above.

invalid or when some activities are no longer required. Changes in resources may be in the amount of a resource required or in the combination of types of resources on any activity. Target dates may be changed and set either earlier or later. Activity durations may change because the activity has been completed, because it has been started but not yet completed, or because the estimate for an activity, not yet started, has been altered.

Changes in logic, resources, or target dates are usually entered on the same type of form as the original input (*Figure 19*) by inserting the revised information. Additionally the previous information must be deleted, usually by inserting on a simpler form the start event, end event, instruction code, and current time estimate. The instruction code will cause the computer to delete the original entry. For changes in duration only, the simpler form can be used with an instruction code causing the revised duration to replace the original duration without another complete entry.

Computer or manual analysis?

The preceding sections have described the input required for simple network analysis programs and the types of output that may be obtained assuming that a computer is to be used. The decision whether or not to use a computer for a particular project will depend both on the size and complexity of the analysis and on the availability and cost of computer time. If the organization has its own computer it may be possible to perform network analyses on it (not all computers are suitable); where a suitable computer is not available within the organization, time can be bought at one of the service bureaux.

Bureaux charges will be an important factor but costing methods found in a survey of computer bureaux were so diverse that it proved impossible to include meaningful figures in the summary of currently available programs given in Appendix 3. However, it should be remembered that service bureaux charges are not the only costs that will be incurred. Equally important is the cost of administering the network by computer, considerable time will be spent in data preparation, time may be lost if forms have to be sent by post, etc. Against these costs must be set the increased reliability of computer results.

No firm figures can be given as to when one ought or ought not to use a computer since costs vary so widely with circumstances. A rule of thumb is that the most elementary analyses can be undertaken satisfactorily by hand for networks of up to two hundred activities, but that as the analytical complexity increases it becomes worth using a computer for smaller networks.

Part III

Further Aspects of the Technique

So far we have considered time-only analysis; in many applications of CPA it will become apparent that this is not sufficient. The most important deficiency is that no account is taken of resource requirements and availabilities. If resources are a problem in only a small part of the network, it may be possible to redraw the network locally to take account of the restrictions. However, for any substantial resource problem a method taking full account of resources is essential. Such methods are considered in the first chapter of this Part.

A difficulty sometimes encountered in projects with a research and development content is uncertainty. Durations may be uncertain and so may the project logic; some activities may, for example, only occur if preceding activities have been successful. A description of the original PERT method of dealing with uncertain durations together with other methods is given in Chapter 8, and probabilistic network logic is considered in Chapter 9.

The last chapter of this Part considers the extension of network techniques to the area of cost control. This should not be confused with cost-optimization techniques, such as CPM, which are disguised resource allocation methods and are briefly described on p. 51. The linking of networks and cost-control techniques arises from the desire to improve on current cost-control methods, and the strength of network methods as a control tool.

The developments of the basic technique are described here, and have so far been used, separately: resource allocation; or probabilistic methods; or cost control. There is no reason in principle why they should not be used together; however, the practical difficulties are enormous. Already projects can be planned using different techniques at different stages; for example, probabilistic times or logic in the early planning and development stages, followed by resource allocation or cost control in the construction stage. Developments of more comprehensive programs in the future may permit further integration of techniques. Meanwhile the technique must be chosen to deal with the most important problem, local intrusion of other problems being dealt with manually.

7 · Resource Allocation*

Introduction

In most organizations with problems of project control, it has been found necessary on many projects to replace straightforward time-only analysis with resource allocation techniques. Resource allocation is used here to cover any network technique which considers resources. A resource is a physical variable, such as labour, space, equipment, material, or finance, the use of which may impose limits on the freedom of a project over time. The need for consideration of resources is most pressing when there are conflicting demands for the same limited quantity and type of resource which may be satisfied in a number of ways; each way may have a profound effect on the overall timing of a project.

For a simple example, consider two similar activities each requiring the services of one skilled man, and only one such man is available. Suppose that the logic says that both activities can be done in parallel and that the network requires them to overlap if the project is to be completed on time. In fact, it can be shown that the delay to the project will be minimal if the activity with the least 'duration plus total float' is carried out first. A problem as simple as this can be solved locally, but with large networks a more systematic method is essential and the use of a computer becomes a necessity.

Philosophy of use

The correct approach should be established by an organization right at the beginning of its work in this field. Resource allocation usually incurs a compromise, and the choice of this compromise must depend very much on the judgement of managers. It is the purpose of a resource allocation system to provide managers with relevant information to assist them in deciding the course of action to take.

* This paper is based on the Resource Allocation Section of the Report of the Working Party on Large Networks by D. J. McLeod and C. Staffurth. However, consideration of papers by Armstrong,[1] Battersby and Carruthers,[2] Hooper,[3] and Pascoe[4] has led to considerable expansions and amendments.—Ed.

43

The aim should be, either to set down a plan to allocate all types of resources, or to accept a limited plan considering only those resources that are
known to be of a critical nature. The objectives of the planning procedure for
the resources that are to be considered must be carefully thought out. Is the
procedure simply to provide information on the resources required if the
activities are scheduled according to the network diagram, or should the
procedure reschedule activities in order to achieve certain objectives? If
the procedure is to reschedule activities, the objectives of this rescheduling
must be thoroughly considered. It has been suggested that much of the
dissatisfaction with resource allocation procedures has arisen, not from
the programs or algorithms used, but from the failure of the organization
to define its objectives clearly.

The simplest objective for rescheduling is to contain resource requirements, at any time, within prescribed limits; this can result in failure to meet
original target dates. More complex objectives are usually concerned with
maintaining smooth resource requirements or with the rate of build-up and
reduction of requirements as the project proceeds. These objectives are usually
associated with problems of labour management. Objectives of this sort can
be pursued with or without the project duration being kept to a minimum.

Many points of detail must be considered in defining resources for the
planning procedure. The availability of a resource, its nature, and the degree
of interchangeability within the resources are important aspects. Availability may be strictly limited or additional resources may be available at
extra costs; availability may vary over time. A difference must be made
between storable and non-storable resources. Non-storable resources are
resources such as labour or machines; if these are left idle for a period then
their use is lost and cannot be recovered. Storable resources, on the other
hand, are resources such as material which can generally be kept until required. This gives rise to the idea of a pool of resources which is being filled
at a certain rate and is being used at a certain rate. There are two limitations:
the pool can only be filled to the limit of its capacity; and the resources cannot be used if the pool is empty. The units of resource within a category can
be considered as homogeneous, although with human beings this assumption
may not be fully justified.

Flexibility in the use of resources is an asset for a project manager. He
rarely has full control of all resources required for his project, a fact which
may affect the sequence of use of a resource. For example, he may require to
use testing facilities that are under the control of another man. In this situation, the project manager may have to adjust his schedule and co-ordinate
his work with that of the testing facility. Where delivery of raw materials is
unreliable, flexibility in resource usage is essential.

All these points will influence the objectives that are set for the planning
procedure. The usual ultimate consideration is overall profit, requiring the
balancing of the importance of costs and project completion time. Some

techniques consider this balance explicitly, as described in the section on cost optimization techniques below; alternatively the balance can be struck by management.

The techniques

The simplest technique which considers resources is known as Resource Accumulation or Aggregation. The resources required by each activity are accumulated for each resource category against a time scale, without any consideration of the availability of resources. From the normal network processing, dates for activities can be obtained, and the usual method is to accumulate against the earliest start dates, although latest start dates can also be used as an alternative. A simple network, with only one type of

Figure 21 *Simple network with resource requirements*

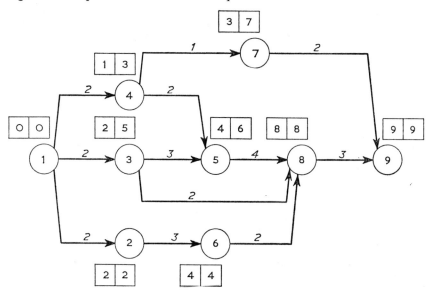

resource, is shown in *Figure 21*; the numbers above the activities are the units of resource required. The resource accumulation in *Figure 22* shows each activity at its earliest start time. The resource requirements can then be compared with availabilities and the plan can be adjusted manually or extra resources can be obtained.

The more complex techniques are those attempting to smooth or level the resource requirements. Nomenclature here is not standardized, but there are basically two approaches according to whether or not the duration must be kept to a minimum. If this minimum is to be maintained, some levelling of resources may still be possible by the use of float, see *Figure 23*. This is sometimes distinguished as resource smoothing. If the minimum project time is not

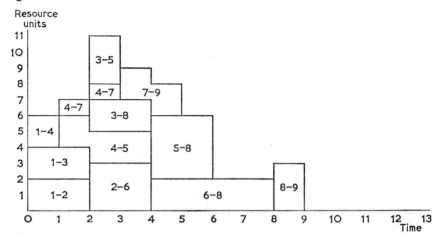

Figure 22   *Resource accumulation*

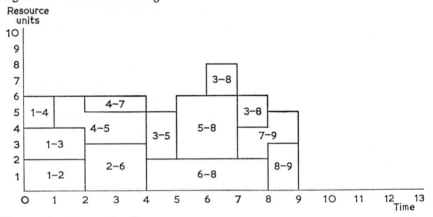

Figure 23   *Resource smoothing*

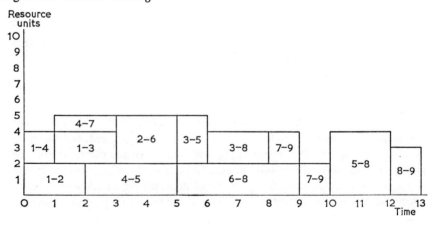

Figure 24   *Resource levelling*

essential, the schedule is rearranged so that a predetermined resource level is never exceeded and, subject to that restriction, project duration is minimized as in *Figure 24*. This is sometimes called resource levelling.

These examples have only considered one type of resource for a small network. In such cases one can investigate by trial and error whether the best solution for the given objectives has been found. However, in practical networks there are usually several types of resource which must be considered and each of many activities may require one or more types of resource. The problem has been to develop methods which give at least good results in practical situations, without excessive computation, although for a network of any significant size a computer is a necessity for resource allocation.

Program methods

The programs in common use employ two basic methods, the Serial Method and the Parallel Method. These are heuristic methods designed to give good, but not necessarily optimal, solutions. There are algorithmic methods, such as Linear Programming, which could be used to give optimal solutions. However these methods would be very expensive because they would require long computing times on a large computer. No attempt has yet been made to apply these algorithmic methods to practical networks.

In the Serial Method, a list is made of all activities in the network, sorted into some order such as total float within latest start date. This list, sometimes known as a Precedence Table, is the basis of allocating the resources. The first activity is considered and the required resources allocated. The second activity is similarly considered, and so on through the list. Should an activity require a resource which is fully employed by an earlier activity, it must wait until the earlier activity has finished with the resource.

The Parallel Method works through the project and for one period at a time selects those activities available for scheduling which can use a given resource. It then sorts the activities into an order of priority. For example, the major sort may be on earliest start date, with a minor sort on total float. Resources are then allocated to those activities with the highest priority. If there are insufficient resources to begin an activity, activities with a lower priority are examined, in case the remaining resources can be employed. If the start of an activity is postponed until the next time period, its earliest start date must be increased and its total float reduced. The network must be examined locally to see whether the dates of the following activities must be adjusted.

Some programs use a combination of these methods and may be called Serial-Parallel. The procedure is exactly as for the Serial Method except that after each activity has been scheduled the precedence order is reviewed and altered, if necessary, in the light of the activity just scheduled.

Every program, whether based on the Serial or the Parallel Method, uses some criteria for determining priorities. For example, the major sort in a Serial program may be by latest start date with total float as a minor sort;

alternatively, latest finish or earliest start could be used for the major sort. Different criteria can similarly be used in Parallel Method programs. Some programs permit the sort of criteria to be specified as program parameters. When considering several resources, the most scarce resource is usually allocated first. If an activity requires several resources simultaneously, it frequently happens that these multiple resources turn out to be the most scarce. Priority to resources is given either arbitrarily or as the result of experience.

Programs also differ in the various minor facilities that are available. Some programs permit activities to be split and scheduled in two or more distinct parts. Some activities, once started, must continue without interruption, others may be started then interrupted and then completed later. This flexibility is exploited to make use of resources that otherwise would be idle. The opposite to splitting activities is the tying of activities. This can occur with, for example, highly specialized plant that needs a continuous feed of activities. The activities may not be on the critical paths of the projects using the facility, but will be actioned in order to keep the special plant running on an economic basis. Some programs also permit activities to have more than one possible resource level and duration. This also adds flexibility to the scheduling process, but should not be confused with the project duration optimization technique described below (p. 51).

Most programs allow resource availability levels to alter from period to period and these levels, together with the normal activity list now including resource requirements, form the main program inputs. Some programs also permit additional premium resources to be used only if target dates are in danger, and some can be set to achieve target dates by breaking resource limits if necessary. Outputs from resource allocation programs vary in detail but are usually variants of two basic forms.

The first type of output is an activity listing, with scheduled start dates, see *Figure 25*. This may be sorted in various ways and may contain other information such as scheduled finish date, delay in starting beyond the earliest start, etc. In the example shown, which represents part of a large network, resource restrictions are not stringent; only one activity, 79–105, has been delayed beyond its earliest start date, and even that is well within the available float. The second type of output is the resource utilization table, see *Figure 26*. For each time period, by period number or by date, the number of units of each type of resource required is recorded; this information is sometimes presented in bar-chart form. A useful derivation from the resource utilization table is the resources remaining table, see *Figure 27*. This indicates the idle resources remaining and draws attention to tight periods with zero or negative resources remaining (in the case where resource limits have been exceeded to maintain target dates) and also to under-utilized resources. A small reservoir of idle resources can be a help in day-to-day management to allow for the effects of bad estimating or sudden plan changes.

Figure 25 *Resource allocation; activity output by scheduled start**

I	J	D	TS	TC	Description	Resources	SS	SF	LF	SD	SS date	SF date	LF date
1037	1041	130			Write Programs	M1	230	359	479		7 Jul. 67	9 Jan. 68	28 Jun. 68
79	105	5			Desn + App A/C Input Forms	L1	232	236	524	50	11 Jul. 67	17 Jul. 67	2 Sep. 68
137	155	5			Test Program K	F1	232	236	514		11 Jul. 67	17 Jul. 67	16 Aug. 68
105	191	20			Print A/C Input Forms		237	256	544		18 Jul. 67	14 Aug. 67	30 Sep. 68
129	165	10			Test Program C	E1	242	251	514		25 Jul. 67	7 Aug. 67	16 Aug. 68
143	159	5			Test Program B	F1	242	246	514		25 Jul. 67	31 Jul. 67	16 Aug. 68
147	165	10			Test Program F	F1	242	251	514		25 Jul. 67	7 Aug. 67	16 Aug. 68
1055	1057	15			Write Clerical Instns. USA		242	256	564		25 Jul. 67	14 Aug. 67	28 Oct. 68
165	167	10			Design Stats. System	E1	252	261	533		8 Aug. 67	21 Aug. 67	13 Sep. 68
165	171	3			Design Account Inquiry System	L1	252	254	538		8 Aug. 67	10 Aug. 67	20 Sep. 68
171	183	1			Program Account Inquiry System	F2	255	255	539		11 Aug. 67	11 Aug. 67	23 Sep. 68
183	191	5			Test Account Inquiry System	F1 I1 L1	256	260	544		14 Aug. 67	18 Aug. 67	30 Sep. 68
67	75	5			Install Computer	J1	257	261	478		15 Aug. 67	21 Aug. 67	27 Jun. 68
75	93	10		478	Computer Acceptance Trials	D1 I2 J1 K1	262	271	514		22 Aug. 67	5 Sep. 67	16 Aug. 68
127	131	0					262		509		22 Aug. 67		9 Aug. 68
127	133	0					262		509		22 Aug. 67		9 Aug. 68
131	155	5			Test Program I	F1	262	266	514		22 Aug. 67	29 Aug. 67	16 Aug. 68
133	155	5			Test Program J	L1	262	266	514		22 Aug. 67	29 Aug. 67	16 Aug. 68

Key I: Start Event Number
J: End Event Number
D: Duration

TS: Target Start
TC: Target Completion
SD: Scheduled Delay

* This output uses the working-day convention described on pp. 14–15 above.

Figure 26 *Resource utilization table*

Period	Resources												
	A	B	C	D	E	F	G	H	I	J	K	L	M
251	0	0	0	0	1	1	0	0	0	1	0	0	1
252	0	0	0	0	1	0	0	0	0	1	0	1	1
253	0	0	0	0	1	0	0	0	0	1	0	1	0
254	0	0	0	0	1	0	0	0	0	1	0	1	1
255	0	0	0	0	1	2	0	0	0	1	0	0	1
256	0	0	0	0	1	1	0	0	1	1	0	1	1
257	0	0	0	0	1	1	0	0	1	1	0	1	1
258	0	0	0	0	1	1	0	0	1	1	0	1	1
259	0	0	0	0	1	1	0	0	1	1	0	1	1
260	0	0	0	0	1	1	0	0	1	1	0	1	1
261	0	0	0	0	1	0	0	0	0	1	0	0	1
262	0	0	0	1	0	1	0	0	2	1	1	1	1
263	0	0	0	1	0	1	0	0	2	1	1	1	1
264	0	0	0	1	0	1	0	0	2	1	1	1	1
265	0	0	0	1	0	1	0	0	2	1	1	1	1
266	0	0	0	1	0	1	0	0	2	1	1	1	1
267	0	0	0	1	0	0	0	0	2	1	1	0	1

Figure 27 *Resources remaining table*

Period	Resources												
	A	B	C	D	E	F	G	H	I	J	K	L	M
251	1	1	1	1	0	2	3	5	2	9	1	1	0
252	1	1	1	1	0	3	3	5	2	9	1	0	0
253	1	1	1	1	0	3	3	5	2	9	1	0	1
254	1	1	1	1	0	3	3	5	2	9	1	0	0
255	1	1	1	1	0	1	3	5	2	9	1	1	0
256	1	1	1	1	0	2	3	5	1	9	1	0	0
257	1	1	1	1	0	2	3	5	1	9	1	0	0
258	1	1	1	1	0	2	3	5	1	9	1	0	0
259	1	1	1	1	0	2	3	5	1	9	1	0	0
260	1	1	1	1	0	2	3	5	1	9	1	0	0
261	1	1	1	1	0	3	3	5	2	9	1	1	0
262	1	1	1	0	1	2	3	5	0	9	0	0	0
263	1	1	1	0	1	2	3	5	0	9	0	0	0
264	1	1	1	0	1	2	3	5	0	9	0	0	0
265	1	1	1	0	1	2	3	5	0	9	0	0	0
266	1	1	1	0	1	2	3	5	0	9	0	0	0
267	1	1	1	0	1	3	3	5	0	9	0	1	0

Multi-project scheduling

In an organization where many projects are running concurrently and drawing on the same resources it may be advisable to schedule them together. If this is attempted additional difficulties arise. These projects are competing for the same resources; extra information is required to determine the relative time scales of each project, and their relative importance. Relative project importance can be defined by giving each project a priority number, based perhaps on the estimated cost of overrunning the end date. Thus the project with the highest priority will be scheduled first, the lower-priority projects following and being delayed if resources are not available. It is questionable whether priorities remain valid over a period of time. Some projects must inevitably become more important while others decline, yet to alter project priorities can lead to confusion. The rule of thumb at present is to leave priorities unchanged once work has started, with a reappraisal when a project is completed or a new one introduced.

A specific problem for managers with several projects running concurrently, is the float consistency between projects. Several weeks' float may be normal and acceptable on one project whereas it would be serious on another. This raises the question of whether float is a good basis for smoothing resources, or as a secondary sort. With single projects it can be satisfactory, but with multi-projects it is doubtful.

Cost-optimization techniques

One of the earlier developments in network techniques was that of overall cost optimization. This approach is to permit each activity to be scheduled at a 'normal' duration and cost or at a 'crash' duration and cost. Sometimes three durations and associated costs are permitted and sometimes continuous linear and non-linear time-cost functions are assumed. An optimization technique is then used to balance the minimum additional cost required to speed up project completion against the gain from earlier completion. The gain may be a reflection of lower overheads or may include an element of opportunity cost. The techniques may also be used to obtain the maximum cost schedule for any given project duration.

This brief description is included in this section because cost optimization is in effect a resource technique in disguise. The different activity durations and costs are a reflection of different levels of resources. However, no considerations of resource availability are involved. These techniques have not been extensively used in Britain and probably their most successful applications have been on small plan-only projects, see p. 31.

Which technique and program?

Resource accumulation has been used very successfully as an aid to project control, usually in conjunction with some manual levelling or smoothing of

Figure 28 *Summary of principal package programs, Spring* 1966

Acronym	GRASP	RPSM	MOSS
Source	IBM	CEIR	CEIR
Name of Program	General Resource Allocation and Scheduling Program	Resource Planning and Scheduling Method	Multi-Operational Scheduling System
Objective	Resource Allocation using fixed resource levels or fixed time period	1. To utilize all resources as smoothly as possible within a set project duration 2. To keep resource levels fixed and see their effect on project duration	To set resource levels or established financial budget Either multi-projects or one large project
Number of Resources	50	26, with maximum of 4 per activity	26, with maximum of 4 per activity
Resource levels	Variable (cyclic or non-cyclic)	Variable (cyclic or non-cyclic)	Variable (cyclic or non-cyclic)
Number of Activities	5000	7500	
Splittability	Facilities for tying and splitting available	Activities can be splittable or non-splittable	
Pool Resources	No	No	No
General	Scheduling priority given by any pair of Early Start, Late Start, Early Finish, Late Finish, Total Float and Duration, Duration and Priority Number	CEIR suggest successive runs to improve the solution	This method is said to be applicable to an unlimited number of projects An extension of CPM and RPSM
Available from Service Bureaux	IBM Data Centre only	Yes	Yes

PREDICT		
ICT	ICT	English Electric Leo Marconi
PERT Resource Expenditure Determination ICT	1900 PERT	Resource Allocation
1. To keep established project date 2. To set levels of resources	Forms an integrated system for time analysis, resource allocation, and cost control Resources allocated under limited resources or limited time	Resources allocated to satisfy stated maximum project delay When the delay is small, the the program is time-limited When delay is very large, the program is effectively resource-limited
24, with maximum of 4 per activity	125, with maximum of 60 per activity	56, with maximum of 7 per activity
Variable to a total of 7 levels, cyclic or non-cyclic	Variable to total of 120 levels, cyclic or non-cyclic Resource may be used for whole or part of activity	Up to 256, non-cyclic
About 2000	6000 (up to 60,000 with skeletons)	Limited to 2000 in the cross section of the network, i.e. those activities considered in current time interval
Splittable activities must be defined Force finish is also available	As PREDICT In addition minimum split may be defined	Activities can have a minimum split, phase split, or be non-splittable Facilities for tying-in tied chains and loosely tied chains
No	Yes	No, two types of resource are available, though: Type A—amount used on each day Type B—amount used in total
Designed to run with ICT PERT on 1500 series machine	Will accept networks punched on cards in the PREDICT format	The method used, although based on the parallel method, is considerably modified Priority given by Latest Start or Latest Finish
Yes	Yes	Yes

resources. Some experience of this method may be advantageous before proceeding to the more complex methods of resource allocation.

Allocation programs based on Serial, Parallel, and Serial-Parallel Methods have all been used on projects with or without duration restrictions. The Serial Method was used more often in the early applications of resource allocation. It requires less computer time and the criteria used are simpler. Satisfactory results have been obtained with this method and a useful approach is to try several initial runs with differing resource and time restrictions to establish the desired balance. Research by Pascoe[4] has indicated that Parallel Methods are superior for many networks and increasing practical experience with programs using Parallel Methods has tended to confirm this. However the criteria used are more complex and it is suggested that only practical experience of a particular program can lead to an understanding of its idiosyncracies.

The different sort criteria used in both Serial and Parallel Methods were also assessed by Pascoe. The only significant difference he found was that latest start and latest finish were the best criteria for major sort; these are both long-term criteria derived by tracing the network through to its final node. However these results have not yet been validated on any substantial practical scale.

The choice between writing one's own resource allocation program or using a package program available on the market must be carefully considered. Writing one's own program has the advantage that the thinking of the planners is kept closely in touch with the capabilities of the computer, the intricacies of program writing, and the answers it produces. One stands a better chance of getting exactly what one requires; however, the difficulties and expense in developing a new and complex program are considerable. The package programs can often be cheaper, but they suffer from the disadvantage that their answers may be accepted without a full understanding of the mechanism by which they are obtained.

Details of the principal package programs available in Britain in spring 1966 are given in *Figure 28*. The summary of programs in Appendix 3 gives brief but more up-to-date details of a wide range of programs. Costs vary widely according to the complexity and size of the network and the outputs required, since many of the bureaux programs are charged on a time and materials basis.

There are many points to be considered in deciding which type of resource allocation technique is the most suitable. It is important that the objectives to be achieved should be carefully considered. The type of project must then be examined. For projects likely to suffer frequent changes in technical content and logic, a less sophisticated method may be preferable compared with a project whose logic is fairly certain to remain constant. This susceptibility to change is often a feature of very large networks if for no other reason than their size.

In choosing between particular programs, factors of cost and accessibility are always important, but so too are the criteria used by the program and the various special facilities offered. These should be measured against the declared objectives of resource planning for the project in hand as far as possible, although it has already been stated that only actual use of any program can lead to a full understanding of the way in which it works.

Monitoring of resource networks

The updating of a network including resources is inevitably more complex than for a straight time network, although the procedure is basically the same. In addition to any changes in logic and durations, changes in resource requirements and availabilities must also be recorded. There are differences of opinion on the frequency and extent of resource updating required. Some firms have found that a complete resource updating, alternating with a time-only updating, is satisfactory.

A minority of the working party on large networks considered that detailed schedules prepared by resource allocation can only be useful for about one to three months into the future. The majority view was that forward-looking information is essential for highlighting potential difficulties, and therefore full processing of a network with resource allocation was preferred. The deliberate suppression of information regarding the future beyond the detailed period of the processing date, to everybody except the project manager and planning group, has advantages. The main advantage is that proper control can be exercised with plan changes that will not confuse functional management whenever a change occurs. This is particularly true at the beginning phase of a project. It is here that management and design skills are at a premium, because these resources are often limited and good co-ordination of these efforts is invaluable.

Conclusions

Resource allocation becomes necessary when conflicting demands are made on scarce resources. Resource accumulation is a method which gives a rough idea of the size of the problem, and manual adjustment of activity dates within their total float can help smooth out excess demands on a resource. Resource allocation programs reschedule resources according to decision rules in an attempt to meet certain objectives.

The choice of technique and program depends on many factors. The effects of the simpler programs are easier to understand, at least until some experience has been obtained; however, there is some evidence that the more complex programs produce better schedules. The type of project is an important factor in the choice of technique. The best answer at present is to have a good resource allocation program with which management is familiar, and for any necessary decisions and adjustments to be made by management in the light of the latest information.

Resource allocation is not an exact science; considerable art is required to apply successfully the techniques outlined in this paper.

REFERENCES

1 D. J. ARMSTRONG (1966), 'Resource levelling and allocation', Internal communication to CPA Study Group.
2 A. BATTERSBY and J. A. CARRUTHERS (1966), 'Advances in Critical Path Methods', *Operational Research Quarterly*, **17**, 4.
3 P. C. HOOPER (1965), 'Resource Allocation and Levelling', *Proceedings of 3rd CPA Symposium*, Operational Research Society, London.
4 T. L. PASCOE (1965), 'Heuristic Methods for Allocating Resources', Ph.D. Thesis, University of Cambridge.

8 · Probabilistic Times

J. A. CARRUTHERS

Introduction

CPA provides a strict framework for a logical sequence of activities, with well-defined durations. Practical projects may have loosely defined sequences of activities, with durations which are difficult to estimate. Since often the main value of CPA is to give a powerful planning tool where none existed previously, these loosenesses in definitions may still be tolerated. However, in many projects, especially large ones co-ordinating many types of sub-contractor with differing degrees of sophistication, potential errors in estimating durations may be significant. The resulting errors in estimating project times and loss of project control may be so considerable as to require a more detailed assessment and treatment of the problem. This is done here; the problems of loosely defined logical sequences, which may become severe in speculative or research and development projects, is treated in Chapter 9.

Estimates of activity durations

The basic problem we are considering arises from the uncertainty in estimating an activity duration and the probable differences between the estimated and the actual duration.

The ideal estimate should be accurate, unbiased, and consistent. Accuracy means that the deviations between estimated and actual duration should be small. By unbiased, one means that the deviations between estimated and actual durations should in the long run tend to zero, positive and negative deviations tending to cancel each other. Consistency means that if estimates were required for exactly the same activity on different occasions but identical circumstances, then identical estimates would be given.

Distribution of errors

Actual estimates will not in general be ideal. Errors, due to inaccuracy, bias, and inconsistency will remain, and one should therefore be able to quantify their effect. The first step is to determine the statistical distribution of errors.

The first American PERT system[1] in 1959 considered the activity durations as variables and used the Beta distribution to represent their distribution. It was not suggested that durations actually followed this distribution, merely that it was a convenient and not unreasonable model to take. The form of the Beta distribution is:

$$f(x) = [(b - a)^{p+q-1}B(p,q)]^{-1}(x - a)^{p-1}(b - x)^{q-1}$$

Where B is the beta function of mathematics and a and b are the finite lower and upper limits to the activity duration. Estimates are thus required for a, described as the most optimistic time for completing the activity, and b, the most pessimistic time. The other two parameters of the distribution are fixed by estimating the most likely time, or mode (m_o), and by assuming that the standard deviation of the distribution is one sixth of the range of possible values. Hence:

$$V = \frac{1}{36}(b - a)^2$$

The mean of the distribution, m, is used as the estimator for the activity duration:

$$m = \frac{1}{6}(a + 4m_o + b)$$

This formula for m is in fact only the true mean for particular values of p and q but is a reasonable approximation for non-extreme values.

The apparent advantages of using the Beta distribution are several. First, the estimator involved in an uncertain activity is clearly given leeway in his estimate, together with a method of quantifying this reasonably. Second, the distribution appears at face value to be statistically sensible, with finite upper and lower limits to activity durations. Third, the calculations to obtain the mean and the variance are simple.

Accepting the first two points, and the fact that the first PERT system was devised in a very short time, the invention and adoption of the PERT system appear reasonable. Since then, numerous articles concerning this particular distribution have appeared (2, 3, 4). Many organizations now use it, with the proviso that 'all three estimates, a, m_o, b be identical'!

The original PERT system was devised for the Polaris missile project, which co-ordinated many different organizations with different qualities of estimate, so that considerable variability might be expected. Simms[5] has described an analysis of the accuracy of estimating within a few large organizations, accustomed to using network analysis with considerable care and expertness, where less variability would be expected. The analysis showed that errors in the activity time-estimates as a fraction of the original estimate had a unimodal, non-rectangular distribution. There was a bias of 25–30 per cent in estimates, and the frequency distribution was of the form:

$$f(x) = Kx^3 e^{-ax}$$

with a standard error of 60–70 per cent; there were still appreciable occurrences in the tails, with 700 per cent errors in estimating!

To conclude this section, the problem of accuracy can be a serious one, even in quite sophisticated organizations. Until an organization has methodically conducted the type of 'feedback' exercise on activity estimates quoted above, the problem of bias could also be serious. These problems, and that of consistency, can be alleviated (though not eliminated) by study, and by careful definition of estimating procedures. An organization would do well to tackle the problem of estimating activity durations by Simms's approach, and not be misled into studying the large volume of papers following up the Beta distribution, which appears to have become academically fashionable.

Estimates of project duration

Having obtained by any means, such as those described in the previous section, distributions for the durations of individual activities, it is necessary to combine these to form a statistical distribution for the duration of a whole network, which is the primary objective of the exercise. Networks are composed of activities in series, in parallel, and in more complex combinations of the two.

Consider first the ith activity in a network; let the probability that its duration lies between t_i and $(t_i + dt_i)$ be $\phi_i(t_i)\, dt_i$. By the usual definitions, we have the mean value and variance of t_i as:

$$\bar{t}_i = \int \phi_i(t_i) t_i \, dt_i$$

$$V(t_i) = \int \phi_i(t_i)(t_i - \bar{t}_i)^2 \, dt_i$$

where the integration is over the range of the activity times.

If a network consists simply of activities $i, j, \ldots k$, in series, the project duration, t_p, is given by:

$$t_p = t_i + t_j + \ldots + t_k$$

and hence

$$\bar{t}_p = \bar{t}_i + \bar{t}_j + \ldots + \bar{t}_k$$
$$V(t_p) = V(t_i) + V(t_j) + \ldots + V(t_k)$$

Thus the mean duration and the variance of a set of activities in series are formed by simple addition – provided that the durations are independent (i.e. that a change in duration for an activity i, due to some factor, does not have a concomitant change in the durations of any of the other activities $j, \ldots k$).

Next consider a network of activities in parallel. With the above notation, let $F_i(t_i)$ be the probability that the duration of activity i does not exceed t_i. Then the project duration t_p, and $F_p(t_p)$ are given by:

$$t_p = \text{Max} (t_i, t_j, \ldots, t_k)$$
$$F_p(t_p) = F_i(t_p).F_j(t_p) \ldots F_k(t_p)$$

There are no simply estimated analytical expressions for $\bar{t}_p$ and $V(t_p)$, as was the case in serial networks, thus:

$$\bar{t}_p = \int_0^\infty [1 - F_i(t)F_j(t) \ldots F_k(t)] \, dt$$

and

$$V(\bar{t}_p) = \int_0^\infty (t - \bar{t})^2 F'_p(t) \, dt$$

The integrals involved can be conveniently evaluated only numerically, or in simplified cases, such as where all F_i have the same analytical form, or all F_i have particularly simple forms.

Practical networks can be regarded as a set of activities in parallel but with crosslinking activities. As a result, the statistical distribution of times in any chain of activities is also affected by the distribution on other chains to which it is crosslinked. The resulting expressions for the statistical distribution of project times become truly formidable and unwieldy. The only way to derive correct distribution of project time is by Monte Carlo methods.[6] One can then attempt some generalizations on the distribution of project times, by inductive reasoning from simple serial or parallel sets of activities, and from Monte Carlo runs on large complex networks with varying degrees of cross-linkage. Moreover, the Monte Carlo approach can cater for the practical situation where durations of activities are statistically correlated.

The PERT approach cuts through the above problems, by blithely disregarding them! In the PERT system, the mean value for the project duration is taken as the sum of the mean values of the durations of the activities on the critical path. This assumption is rigorously correct for a project which is a single chain of activities, but progressively underestimates the mean project duration as the number of parallel activities, or crosslinkages, increases. The variance for the project duration is taken as the sum of the variances of the activities on the critical. Again, this assumption is rigorously correct for a project of serial activities, but progressively underestimates the variance as the degree of crosslinkage increases.

From these assumptions, the PERT system then quotes the mean project duration, and also its cumulative distribution function $F(t)$, so that one can read off the percentage probability, $F(t)$, that the project duration will be less than a specified amount, t. For the mean duration m, $F(m) = 50$ per cent.

In spite of the invalidity of combining distributions of activity durations, and in spite of the dubious relevance of the Beta distribution, the PERT system of quoting probabilities for completion dates can be useful to the management of projects with highly uncertain durations. Another advantage is in highlighting 'subcritical paths', that is, paths whose durations are less than the critical path, but whose variance is sufficiently large to give them a high probability of being critical. Nevertheless, a simpler management expedient is to progress all paths whose durations are within a few per cent of that of the critical path.

REFERENCES

1 D. G. MALCOLM, J. H. ROSEBOOM, C. E. CLARK, and W. FAZAR (1959), 'Application of a Technique for Research and Development Program Evaluation', *Opns. Res.*, **7**, 646–69.
2 C. E. CLARK (1962), 'The PERT Model for the Distribution of an Activity Time', *Opns. Res.*, **10**, 405–6.
3 F. E. GRUBBS (1962), 'Attempts to Validate Certain PERT Statistics', *Opns. Res.*, **10**, 912–15.
4 W. A. DONALDSON (1965), 'The Estimation of the Mean and Variance of a "PERT" Activity Time', *Opns. Res.*, **13**, 382–5.
5 A. G. SIMMS (1965), 'Some Thoughts on the Mechanism of Estimating', Internal Communication to CPA Study Group, OR Society, London.
6 R. M. VAN SLYKE (1965), 'Monte Carlo Methods and the PERT Problem', *Opns. Res.*, **13**, 839–60.

9 · Probabilistic Networks

C. STAFFURTH

Introduction

After network analysis techniques such as PERT and CPM were established, it was pointed out by Eisner[1] that to assume that each activity would be successful was unnecessarily restrictive. A more general view is to consider that activities will not necessarily be successful, in which case alternative courses of action should be considered.

For example, in *Figure 29*, activity C may follow activity A if A is successful, otherwise activity B will follow A. Thus a probability can be assigned to the choice of B or C following A, based on the outcome of A.

Figure 29   *Probabilistic activities*

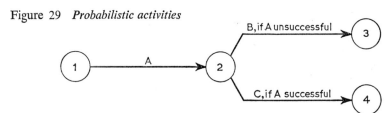

Following Eisner, Elmaghraby[2] presented an algebra for dealing with the probabilities assigned to activities in a network, and a notation for showing the possible courses of action on a network diagram.

These techniques were received sympathetically, because CPA could not illustrate satisfactorily the design–test–modify–retest cycle that often occurs in research projects. Nevertheless, it was recognized that the existence of a technique was of little interest unless it could be applied with advantage to projects already planned and controlled by CPA. Indeed the concept of probabilities has not been used so widely as the basic network techniques, based on certainty.

This paper reviews the subject and looks at areas to which it has been applied successfully.

Notation and algebra

The node in a probabilistic network (corresponding to an event in a deterministic network) may be thought of as a decision box, and as consisting of two halves: a 'receiving' half and an 'emitting' half. An information signal is 'received' at the node and 'emitted' from it according to the logic of the activities entering it and leaving it.

There are three possibilities for each of the 'receiving' and 'emitting' halves; these are shown in *Figure 30*.

Figure 30   *Nodes as decision boxes*

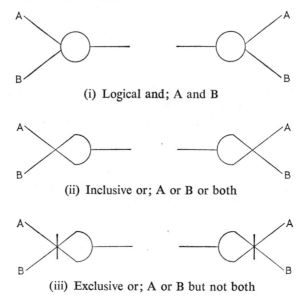

(i) Logical and; A and B

(ii) Inclusive or; A or B or both

(iii) Exclusive or; A or B but not both

Briefly the main algebraic results of these possibilities are as follows:

Two activities in series, having estimated times t_a and t_b, and probabilities of occurrence of P_a and P_b, may be replaced by an equivalent activity having an estimated time t_e and probability P_e where

$$t_e = t_a + t_b$$

and

$$P_e = P_a.P_b$$

Two activities in parallel with the *and* relationship, may be replaced by an equivalent activity, where

$$t_e = \text{Max}(t_a, t_b)$$

and

$$P_e = P_a.P_b$$

Two activities in parallel with the *inclusive or* relationship may be replaced by an equivalent activity, where

$$t_e = t_a.P_a + t_b.P_b + (\text{Min}(t_a, t_b) - t_a - t_b)\,P_a.P_b$$

and

$$P_e = P_a + P_b - P_a.P_b$$

Two activities in parallel with the *exclusive or* relationship may be replaced by an equivalent activity, where

$$t_e = t_a.P_a + t_b.P_b$$

and

$$P_e = P_a + P_b - 2P_a.P_b$$

These results are stated baldly, merely to show how the probabilities in a network can be combined; fuller details are given in reference [2]. It is also possible to use simulation techniques to ascertain the probability of achieving certain outcomes. This is discussed further below.

Types of network

'Probabilistic networks' is a high-sounding and esoteric title. Properly, probabilistic networks are those containing alternative activities with probabilities assigned to them. A more informal type of network is called a branched network, containing alternative activities, but without probabilities being assigned to them. Probabilistic and branched networks have one start event and usually only one terminal event.

There is a close association between the theory of probabilistic networks and decision theory. In decision theory one talks about states of nature and pay-offs and outcomes, and the outcomes are usually simple. If the outcomes are complex and require a probabilistic network to describe them, the theory of probabilistic networks might be considered to be a branch of decision theory. A network containing one start event, with many alternative activities and many terminal events, is usually known as a decision tree.

Thus the boundaries of the subject are not clear cut but merge into other fields, particularly decision theory, and, as will be shown below, into investment analysis.

Assignment of probabilities

Probabilities may be ascertained from one of three sources. First, the probabilities of alternative courses of action may be determined from probability theory, decision theory, or game theory. Second, it may be possible to base probabilities upon the statistics of the data of past events. In both these cases, the probability of the terminal events occurring will have a sound foundation.

The third source is the subjective judgement of the person most closely associated with the activities concerned. It is usually sufficient to consider only probabilities of $0.9/0.1$, $0.8/0.2$, $0.7/0.3$, $0.6/0.4$, or $0.5/0.5$. These are quite good enough to give useful indicative results, especially if there are many alternative activities in the network. A good exercise is to compute the overall probability after reversing the probabilities of individual alternatives. It will be found that the outcome is scarcely affected by some decisions, but will be very sensitive to others; it then pays to re-assess those probabilities more carefully.

Results

A probabilistic network leading to a single terminal event is really a superposition of many networks. It is possible to calculate the shortest path through each of them and the costs involved. In theory one can attribute a vector of parameters to each activity, but the most relevant parameters are probability, time, and costs. By using the algebra of probabilistic networks, one can calculate the expected time to completion of the project, together with an expected cost.

Decisions may be made locally; for example, one may have to await the outcome of a test before deciding which path to take: or decisions may be taken before the project begins, that is to say, the network will show the alternative ways of dealing with the project.

The principal aid to management is to show where to apply additional resources (if they are available) in the form of 'insurance schemes' or certain additional activities in parallel to raise the probability of the desired outcome. This is true whether the object is minimum time, minimum cost, or one particular outcome out of many terminal events. Paths involving these minima frequently have the least probability of occurrence. Additional activities will involve additional cost, and where the objective is minimum cost, the differential between the two paths of lowest cost will determine whether it is worth providing these additional activities.

Examination of a single project with many terminal events, and with decisions to be made before the project starts, is very similar to choosing research projects in which to invest the company's resources. This can be extended to general investment analysis. Investment analysis considers the pay-offs from various projects and assigns to them some figure of merit, such as the rate of return given by Discounted Cash Flow. It is concerned with the flow of capital invested in a project, and the flow of revenue emanating from it. Adelson[5] studies this field, and comes back to decision theory!

Costs of activities may be unique, or may be a function of time. Chilcott[3] finds in expenditure on research projects that there is a general background cost which is a function of time, and in addition alternative paths require different plant or equipment costing considerable amounts of money. He therefore uses the notion of 'gate money' at each decision node, being the fixed capital cost, or setting-up cost, required before a given path can be followed. By extending the algebra to include this gate money, he can relate time, background costs, and total gate money for the whole network to the probability of the occurrence of various outcomes.

In cases where the number of alternatives is of the order of 10^1, analytical methods are suitable. But if the number is of the order of say 10^6 (produced for example by 20 decision nodes each with two alternatives), simulation is required to evaluate the outcomes, as discussed below.

Applications

The basic network analysis techniques have been applied most commonly to construction projects. Although complete certainty may not obtain, alternative paths do not need to be considered, and the assumption of successful outcome is adequate.

Probabilistic networks might be applied to the major overhauls of complex plant, where perhaps it will not be known whether a component needs to be replaced or repaired until the plant has been dismantled. Alternative paths would result from the replacement, repair, or the leaving alone of components. Probabilities could be derived from past experience. However, no such application has been reported.

In research and development projects, where the accent is on development, and the project employs a team of hundreds and is expressed by a large network, considerable uncertainty may exist. Although, if the feasibility of the project has already been demonstrated, the uncertainty is not likely to affect vital issues.

The application of probabilistic networks may not be successful for two reasons. In the first place, it is difficult to pinpoint some areas of uncertainty, and hence define alternative courses of action. Instead, success or certainty is assumed, and it is accepted that considerable corrections may be made to the network when development does not proceed according to plan. Second, the analyst may have an uphill task in selling the results of a subtle piece of analysis to a project manager who controls a very large team involving many skills, and having by necessity a very wide field of view. Furthermore, the project manager must be able to come to a decision that is significantly different from the decision he would have made had he considered merely the critical path of a network based on certainty, as well as the intangible factors not amenable to analysis; otherwise the sophisticated analysis is not worth the effort of carrying it out.

On the other hand, in research projects where the accent may be on pure research or applied research, the project team may consist of between two and twenty people. The project leader himself may draw up a small network describing the project, and process it himself. In this case, probabilistic or branched networks have been shown to give advantages.[4] Provided the project leader can submit himself to the discipline of thinking out the consequences of uncertain activities, the relevant times, costs, and resources required by alternative courses of action, then his mind becomes very clear about the difficulties and challenges of his project. In addition, communication between project leaders and the management has been found to be greatly improved, especially where the project leaders have to bid for scarce resources.

Probabilistic networks have been applied to a different field, namely, the assessment of the reliability of a complex go, no-go system, one in which some components are duplicated or triplicated and interconnected.

The logic of the functioning of the system is expressed as a directed network, in which the arcs represent components which may be unreliable, and the nodes represent interfaces between them. An example of a node with the *inclusive or* relationship is the case where component 1 is the duplicate of component 2, and part of the logic may state that component 3 cannot start to function unless either component 1 or component 2 or both have functioned. The probability of success, i.e. the reliability, of the components is derived from tests or past experience.

A computer is used to employ the techniques of simulation to produce a sample set of component fail/success patterns. If the computer can find a valid path through the network using this pattern, the system is said to be successful, otherwise it has failed. Repeating this simulation many times allows the system reliability to be evaluated.

Conclusions

It was said, in the early days of CPA, that it caused the planners to think more deeply. The successful application of probabilistic networks enforces an even stricter discipline, but gives a much clearer insight into the project concerned.

Its most successful application has been in research projects, and the technique can be used to show where back-up facilities should be brought in to attain the desired probability of success. It also forms a useful guide to the management of a large number of research projects and shows where it is worth investing money and resources. It can also be used in investment analysis where the sequence of events is complicated (but not too complicated).

The two factors common to the successful application of probabilistic networks are: it must be possible to define the alternative course of action, and the number of these alternatives should not greatly exceed two. In addition it is an advantage if the project is expressed as a small network (say 50 activities), allowing manual processing.

A form of probabilistic network has been applied successfully to the assessment of the reliability of complex systems.

REFERENCES

1 H. EISNER (1962), 'A Generalised Network Approach to the Planning and Scheduling of a Research Project', *Opns. Res.*, **10**, 1.

2 s. e. elmaghraby (1964), 'An Algebra for the Analysis of Generalised Activity Networks', *Man. Sci.*, **10**, 3.
3 j. f. chilcott (1965), 'Probability Networks: their Application to Budgeting for Research', Symposium on the Application of Networks to Research Planning, Operational Research Society.
4 r. l. brown (1965), 'Strategies and Schedules', Symposium on the Application of Networks to Research Planning, Operational Research Society.
5 r. m. adelson (1965), 'Criteria for capital investment: an approach through decision theory', *Operational Research Quarterly*, **16**, 1.

10 · Network-based Cost Control

P. M. USHER AND D. W. FORSYTH

Introduction

Network analysis is now firmly established as a major tool in the planning and control of project work, often leading to shorter project durations, fewer delays, and better utilization of available resources. These benefits derive primarily from the method of recording the logical interrelationships between the activities and the subsequent calculation of float or spare time associated with each activity.

In the first instance, however, the emphasis has been largely on physical progress, and while this is certainly important, planning and control can only be complete when physical progress is linked with cost. It may be only too easy, for example, to keep a project to time by working unplanned overtime and exceeding the budget. Traditionally the link between physical progress and cost has often been loose, with the result that a reliable indication of whether or not a project is overrunning budget has been obtained only when it is too late for corrective action to be taken.

Principles of project cost control

If a budgeted cost is allocated to each activity in a work programme, it is possible to spread the costs in time to form an initial cost programme, indicating the rate at which capital will be required throughout the project. This is the conventional S-curve. When the work programme is updated, the budgeted costs of activities completed can be summed to produce the budgeted value of work done. This enables two comparisons to be made:

(*i*) Cost incurred against initial cost programme.
(*ii*) Cost incurred against budgeted value of work done.

The first comparison indicates whether capital is being used at the expected rate and provides a control of cash flow, whereas the second comparison indicates whether work is being done within the budgeted cost. If no further deviations from budget were to occur, this comparison would also represent the final overspend or underspend.

However, when the programme is updated it is possible to revise the estimated costs of future activities, just as it is possible to revise future time

Figure 31   *Cost curves*

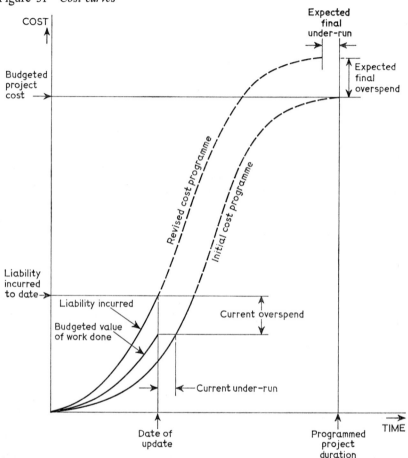

estimates, and to respread these in accordance with the updated programme. By adding these to the cost incurred it is possible to produce a revised cost programme indicating the latest expected rate of requirement of capital and enabling a third comparison (see *Figure 31*) to be made:

   (*iii*) Expected final cost against budgeted final cost.

Network basis

In theory the principles of project cost control can be applied in conjunction with any system of work-programming. In practice there are a number of

reasons why this has not generally been done effectively prior to the introduction of network analysis:

(*i*) Networks provide a more rigorous and detailed work-time relationship and hence allow the preparation of a more reliable cost programme than has been possible hitherto.

(*ii*) The detailed work-time relationship facilitates the accurate assessment
of work done and avoids the subjective estimates of percentage completion so commonly associated with project work and so often linked,
either consciously or subconsciously, with the percentage of total
time or budget expended.

(*iii*) Networks provide an arithmetical basis for handling duration and time
data, thereby facilitating the spreading of costs in time and allowing
an integrated system of cost and progress control to be processed
by computer.

In the simplest case, costs are allocated to individual activities and are
reported against the project as a whole, but such a system is inflexible and of
limited use in practice.

Charge numbers and work packages

In order to provide a flexible system of cost and progress reporting, it is
necessary to break the project down by a system of charge numbers (*Figure
32*). This breakdown, on an hierarchical or explosion basis, should be undertaken at the inception of the project and should form the framework, or
cost plan, against which the budget estimate is prepared and against which
both cost and physical progress will subsequently be planned and controlled.

Figure 32   *Charge numbers*

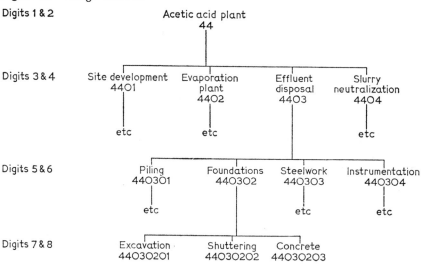

The form of the breakdown may be varied to suit individual projects but, where a number of projects of a similar type are undertaken, it may be helpful to have a list of standard elements and charge numbers which can be used to build up the cost plan for any project.

The unit of input for introducing cost data into the network is the work package, which is linked to the network by means of start and finish nodes which form a normal part of the network. A work package may vary in size from a single activity to an entire network, but, in general, the smaller the size of the work package, the more accurate will be the cost control, and clearly a work package corresponding to an entire network is of little value. Every work package should correspond to an element in the cost plan and preferably, though not necessarily, be taken from a single level of the charge number breakdown.

Within a work package it is necessary to make some estimate of the rate at which costs will be incurred. This is normally achieved by dividing the work package into a number of equal time intervals and estimating the cost increment corresponding to each division. Because of the errors inherent in estimating these increments, the lowest level for cost-reporting is normally at least one level above the work packages. By reporting costs at a level where the charge number comprises a number of work packages, the significance of the errors is reduced for two reasons. First, the errors over a number of work packages tend to cancel out and, second, the errors form a smaller proportion of the reported cost.

Liability for payment

If direct labour is used to execute a project, payments will be very closely related to work done, and budgeted value of work done can be compared directly with the payments made. In other cases, where work is contracted out there is likely to be an interval of time between the liability arising from the completion of an item of work and the actual payment. Because of the variable nature of this interval, it is normally necessary to control on liability which is directly tied to physical work, rather than on payment. At any time the liability will be made up of three elements; payments made, invoices received but not yet paid, and work completed but not yet invoiced.

The value of payments made and invoices received is obtainable from straightforward accounting procedure. The value of work completed but not yet invoiced must be based on estimated or budgeted costs, and its assessment depends on invoices being identifiable with specific points on the network. Invoices must therefore be linked to the network by the date on which the corresponding liability arose, rather than by the date of submission of the invoice.

Limitations

Any form of control is clearly worthwhile only where the information produced can result in effective action. This means that network-based cost control is most useful where the scope for varying the rate and manner in which the various parts of a project are undertaken is greatest.

Where work is contracted out, the value of network-based cost control is reduced until, in the extreme where an entire project is let as a single turnkey contract, network-based cost control is essentially of use only to the contractor. However, if at the start of a contract a cost programme is agreed with the contractor, either on the basis of increments within a single work package, or preferably by breaking down the overall price into a number of work packages, network-based cost control can provide a simple and unarguable basis for progress payment. Otherwise, where a project is let as a number of fixed-price contracts, the value of network-based cost control is largely dependent on the following factors:

(*i*) Number of separate contracts and their interdependence.
(*ii*) Time intervals between letting of contracts.
(*iii*) Availability of capital and importance of cash flow.
(*iv*) Number of reporting levels required.

In certain types of project, as for example in the maintenance of major capital equipment in process plants, the cost of doing the work may be insignificant in relation to the value of having the work completed, and clearly, in such cases, any form of cost control is of limited value.

Reporting and the use of computers

Because of the amount of recording, adding, and sorting necessary to apply network-based cost control in any detail, it is normal to use a computer. One of the advantages of this is the wide variety of reporting information which the sorting facilities of the computer permit. A typical program can, for example, sort on any digit or series of digits within an eighteen-digit charge number.

Normally, reports in terms of both physical progress and cost will be prepared for selected levels of charge number breakdown above work-package level, as appropriate for different levels of management. Further reports by, for example, departmental responsibility, trade, or class of expenditure can be obtained by utilizing one or more of the otherwise unused charge number digits as additional sorting codes. The precise form of the reports will depend on the nature of the project and, in particular, on the degree to which the work has been contracted out. In general the following types of report may be obtained:

Direct work: by project; by work item (at selected levels of charge number breakdown; by trade (*Figure 33*); by location.

Contracted work: by project; by contract (*Figure 34*); by agreed breakdown within contracts.

One of the side-benefits of the sorting facilities of the computer is that information can be produced in a form suitable for feedback to the estimating function (*Figure 35*). In the past, the work involved in analysing a completed project to provide useful feedback has been so great that it has normally been omitted.

Figure 33   *Cost report by trades*

Contract no. 135	Cost report by trades Expected final costs only					Date 4.10.67
	Labour cost		Plant cost		Materials cost	
Trade	Budget	Expected variance	Budget	Expected variance	Budget	Expected variance
Concretor	8,650	−350	3,700	+200	9,420	+100
Plumber etc.	4,700	+310	150	NIL	5,120	+500

Figure 34   *Cost report by contract*

Project No. 47	Progress and cost report by contract							Date 3.5.67
Contract	Progress			Cost to date		Final cost		
No. Description	Progd. compl.	Expected compl.	Remaining float	Budget	Variance	Budget	Expected variance	
23 Piling	3.3.68	7.3.68	+2	370	+30	2,350	+300	
27 Cooling tower etc.	17.5.68	24.5.68	−1	430	−20	17,800	+500	

Figure 35   *Final feedback cost report*

Project No. 403	Estimating feedback report						Date 3.7.67
	Duration		Labour cost		Plant cost		
Activity	Progd.	Final variance	Budget	Final variance	Budget	Final variance	
Control room piling etc.	8	+2	350	NIL	600	+130	

A particular application

In a particular application, a network of the order of 8,000 activities representing work valued at £11 million was broken down into 269 work packages, representing some 30 activities per work package. The analysis was carried out on an IBM 7094 computer using the IBM PERT-Cost program.

A list of the work packages was prepared, showing for each one: start and finish node numbers, description, charge number reference, and estimated cost of the work spread over ten equal time increments. This information was then punched on cards and fed to the computer. A second type of input

Figure 36 *Budgeted liability for work done*

Rescheduled cost programme Non-cumulative		Date 1.9.67
Month	Charge no.	Budgeted liability
06/67	7003	45,000
	7007	7,826
	7503	48,000
	Monthly total	100,826
07/67	7001	64,000
	7004	6,800
etc.		

information gave details of budget cost commitments which were fixed in time irrespective of physical progress. This was fed to the computer in the same way and, together, the total information enabled the computer to prepare the cost programme.

A cost update was carried out at every alternate time update. On the basis of the latest progress information, the computer was able to reschedule the cost programme to produce the budgeted value of work done to the date of reporting (*Figure 36*). At the same time, actual liability incurred for each work package up to the same date was totalled manually and compared with the budgeted value obtained from the computer.

Although, in this instance, a part of the operation was carried out manually, it would nevertheless have been possible for the whole operation to be handled by computer if desired.

Conclusion

As the complexity and tempo of project work increases, so does the need for an integrated system of planning and control of physical progress and cost. Network analysis, combined with the use of computers, provides a valuable step towards achieving this.

Part IV

Large Networks

As an organization's experience of CPA increases, the projects tackled are likely to grow in size and, in certain organizations and industries, may become very large indeed. Very large networks intensify some of the problems associated with smaller networks and also create new problems.

The Large Network Working Party of the CPA Study Group was formed in November 1964 to consider the special problems arising in handling large networks. It defined large networks as those above about 3,000 activities when handling straight time calculations, and above about 1,000 activities when dealing with resource allocation or cost control.

A report was presented to the study group in October 1966 consisting of seven sections, six of which are presented here; the seventh, on resource allocation, has been included in Part III. The first chapter of this Part outlines types of project where large networks may be expected and the special problems likely to be encountered; the subsequent chapters discuss particular problems and methods of solution.

The conclusions of the working party were:

(*i*) The main objective in using large networks is not only to plan logically the work in major projects but to control progress thereafter.

(*ii*) Methods vary according to the nature of the project, the organization involved, and individual preferences.

(*iii*) Success depends on the full support and involvement of top management, and the degree of co-operation which can be achieved between client, consultant, and contractor.

(*iv*) Experience and knowledge of the technique are still developing.

11 · *Major Problem Areas*

E. L. BUESNEL

Introduction

This paper sets out the areas of application which present problems for users of large networks. The form of analysis is by each of eight main types of project, subsectioned by seven technique aspects. While many of the comments apply to most, or all, project types (and even to small projects), they are considered especially relevant where given.

Some generalized comments applicable to any project type are given below.

(*i*) *Network Preparation* – The degree of detail must be geared to the complexity of the project and held to manageable proportions; experience is the best guide. Hierarchical networks are rarely realistic. A library of standard subnets based on repetitive patterns is a useful aid.

(*ii*) *Monitoring* – Milestones increase the effectiveness of reporting progress. Fast-moving activities at site need special attention for control.

(*iii*) *Communications* – The discipline imposed by a network improves communication. Dispersed sites and the many different functions involved are the main difficulties.

(*iv*) *Organization* – Unifying control authority (one man) has advantages and a joint planning approach increases effective usage. The technique can provide good control despite unsuitable organization structure. A policy ruling on usage and top-level support is essential, otherwise much additional effort is needed to get the system going. Applications always raise organization problems, but do not solve them.

(*v*) *Resource Allocation* – Simple computer resource allocation programs plus good management are more effective than versatile but complex resource allocation programs plus poor management. Over-sophistication must be avoided and is seldom justified at present.

(*vi*) *Education and Training* – Senior management need education in CPA to accept the politically unpalatable results which sometimes emerge. Everyone must understand the purpose of detailed control principles behind the technique.

81

(*vii*) *Cost Control* – There is insufficient linkage between accounting procedures. Outdated cost-control methods limit the effectiveness of time control. Cost control is easier in the early stages and increasingly difficult towards the end of projects.

Research and development engineering

Typical examples of this type of project are space research, military weapon systems, and large-scale technical equipment (hovercraft, for example).

(*i*) *Network Preparation* – Sequences are difficult to define and results of activities are often uncertain. A network is often used as a strategic planning method and is sometimes supplemented by branching networks and decision probabilities. The closer definition of work involved is useful to scientists. Design work is difficult to incorporate.

(*ii*) *Monitoring* – Progressing is difficult in the conventional way owing to recycling. 'Success outcome' may be more pertinent for this work. Large numbers of network modifications occur. Stage control (as in the original PERT) is more applicable than activity monitoring. A suitable communication and co-ordination structure is essential. Projects are likely to drift, therefore one must persist with monitoring in order continuously to reset the target.

(*iii*) *Communications* – Understanding between professional interests is often low. Geographical dispersion and highly specialized work mean that information flow is bad. Work identity versus costs and progress must be clearly defined. Large networks are difficult to handle, so use abbreviated diagrams of critical and subcritical paths.

(*iv*) *Organization* – A central co-ordinating office helps with network discipline. Project orientation is essential on large major single projects. In those research and development laboratories where multi-projects are handled, the organization is often departmental, which tends to reflect on communications and monitoring.

(*v*) *Resource Allocation* – Supporting services (e.g. drawing offices, clerical, manual, and analytical) may be a limitation. Priority loading of major resources is important. Non-repetition creates estimating difficulties. Current computer programs are not always suitable. Where probabilistic networks are used, resource allocation becomes difficult. The standard resource allocation programs can be used, provided one is prepared to re-run frequently.

(*vi*) *Education and Training* – Abnormal resistance to network representation requires special 'selling'. Scientists are unwilling to plan although exposed to scientific techniques. Much is yet to be learned about adapting network techniques for many possible alternative sequences.

(*vii*) *Cost Control* – Cost control should aim to estimate total cost with early reporting of significant cost divergence. The choice of path to follow may be determined by how much it costs to enter a new sequence. Difficulties of application arise because historical cost data are not available or experience is insufficient.

Major public undertaking on dispersed site (irregularly occurring)

Typical examples of this type of project are railway systems and services distribution schemes (electricity grid or gas pipeline, for example).

(*i*) *Network Preparation* – Networks may be large, involving many major contractors, who should prepare their own networks. Initial use of a planning consultant is desirable. Diverse responsibilities for planning makes correlation of networks difficult.

(*ii*) *Monitoring* – Data-linking systems are advantageous where geographical problems exist. Alternatively, network analysts must visit construction sites frequently. Updates usually involve logic changes, and the preparation of revised schedules can take several weeks (which may mean six to eight weeks between updates).

(*iii*) *Communications* – Understanding between functional interests is often low. A master network can clear communications hurdles. Communications between client and contractors can be maintained through a resident engineer well briefed in overall programme implications.

(*iv*) *Organization* – Functional organization is often unavoidable. A client's central planning team is of particular value despite the possible absence of real authority. Project organization is unique and probably non-repeating.

(*v*) *Resource Allocation* – Contractors are responsible for their own resource allocation. For some projects a national availability of resources may need to be assessed, e.g. tunnel miners. Efficient organization of men and equipment can be achieved, but rationalization between contractors is usually very difficult.

(*vi*) *Education and Training* – Seminars and courses are required on a major scale. Uniformity of training standards is difficult to obtain. Top levels are not always convinced about, or available for, training.

(*vii*) *Cost Control* – The rate of expenditure is the best guide to rate of progress. Networks are important for cost/budget prediction. PERT/COST ideas are particularly appropriate and useful as means of bringing together multiple contractors' costing. Project cost control demands major revisions to accountancy procedures.

Major public undertaking on localized site (regularly recurring)

Typical examples of this type of project are power station and ship construction.

(*i*) *Network Preparation* – Networks may be very large involving many subcontractors preparing their own subnetworks. CEGB use a central projects office as the source of networking skills.

(*ii*) *Monitoring* – Data-linking systems are advantageous where geographical problems exist. Alternatively, network analysts must visit construction sites frequently. Updates usually involve logic changes and the preparation of revised schedules can take several weeks (which may mean six to eight weeks between updates). Progress information is not easily obtained for overall control of different contracts.

(*iii*) *Communication* – Understanding between functional interests is often low. A master network can clear communication hurdles. Communications between client and contractors can be maintained through a resident engineer well briefed in overall programme implications; this also applies to manufacturers. There are difficulties in defining detail design and relating particular drawings and specifications to an activity. Use of the same contractors each time makes cumulative planning expertise available for the next job.

(*iv*) *Organization* – A project manager and a strong central office are essential (as in CEGB or consortium). Contractual agreement for submission of networks and joint planning is essential.

(*v*) *Resource Allocation* – Resource allocation is not normally a problem at client level but individual contractors will need to implement it. Pre-contract work, design, and drawing offices are areas requiring resource allocation.

(*vi*) *Education and Training* – The central client office usually has experience and sets the standard of training requirements and can run courses for contractors. Top levels of organizations concerned are not always convinced of the value of courses, etc., or the basic psychology of participation.

(*vii*) *Cost Control* – The rate of expenditure is the best guide to rate of progress. Networks are important for cost/budget prediction. PERT/COST ideas are particularly appropriate and useful as means of bringing together multiple contractors' costing. Project cost control demands major revisions to accountancy procedures.

Capital plant for manufacturing industry

Typical examples of this type of project are the construction of a chemical plant complex, or an oil refinery.

(*i*) *Network Preparation* – Networks are large but relatively simple. Extended use of standard subnetworks for plant installation speeds up the process of diagramming.

(*ii*) *Monitoring* – If the work is mainly departmental, monitoring can be tight. Frequency of update is increased (i.e. weekly) at shop floor and on site. It is difficult to avoid 'playing by ear' when delivery delays are frequent.

(*iii*) *Communications* – Foremen and chargehands accept the technique slowly. Interdepartmental boundaries may be high and careful handling of information is essential for first project applications. Communications are good, owing to tight organization.

(*iv*) *Organization* – A strong central planning office is needed as a service to all departments. Organization must be project-oriented. Reporting and assessment to top management are conveniently carried out by central planning.

(*v*) *Resource Allocation* – Design and drawing office work and engineering specialists are usually a limiting factor, as also are the 'administration' resources. An additional problem is the overlap of projects (i.e. multi-project scheduling and priority factors).

(*vi*) *Education and Training* – Training courses can be localized within the company but need to be designed for different levels of management. Refresher courses are also useful to renew people's contact with the technique and to introduce advancing methods.

(*vii*) *Cost Control* – The rate of expenditure is the best guide to rate of progress. Networks are important for cost/budget prediction. PERT/COST ideas are particularly appropriate and useful as means of bringing together multiple contractors' costing. Project cost control demands major revisions to accountancy procedures. There is a vital need for cost prediction and continuous re-forecasting of 'cost at completion'. Use of capital at site is a measure of rate of return. The interest factor is not usually taken into account.

New product introduction (including marketing and distribution)
Typical examples are a new range of computers, consumer durables, and a new range of motor-cars up to production.

(*i*) *Network Preparation* – Considerable manipulation is necessary to meet fixed launch dates. Logic may be diffuse, imprecise, and very changeable. Large networks may be undesirable. Often 'plan-only' networks are useful to clarify the approach.

(*ii*) *Monitoring* – Control is difficult because of varied functional involvement. Precise reporting is not easy, owing to lack of clarity in activity content. Complete redrawing of networks is often necessary.

(*iii*) *Communications* – Existing functional organization difficulties loom large. 'Language' problems exist between marketing and technical people.

(*iv*) *Organization* – Existing functional organization difficulties loom large. A project manager and a co-ordinating specialist planner are essential.

(*v*) *Resource Allocation* – This is applicable for design, drawing office, and tooling-up work-load but is very difficult for the marketing and administrative staff involved. Probably forward global assessments of departmental loading are desirable.

(*vi*) *Education and Training* – There are extreme problems here because of the different functions and level of personnel involved. Requirements and timing of training are difficult. Top levels are not always convinced of the value of the technique.

(*vii*) *Cost Control* – Opportunity pay-off is feasible for meeting a deadline.

One-off building construction
Typical examples of this are the construction of a hospital, office block, library, or university complex.

C P A—G

(*i*) *Network Preparation* – Client complexities may mean incomplete networks and lack of full information at the right time. There is usually a need for at least two correlated networks; the client's and the contractor's. Lack of planning by the client often defeats the contractor's planning. The latter must also accommodate changes due to weather and soil conditions. Administration logic is difficult to state. Supervision is extremely difficult to maintain.

(*ii*) *Monitoring* – Many subcontractors fail to give accurate progress data. Minor changes can usually be handled at site but major revisions must be dealt with at headquarters. Weekly updates are desirable. 'Stage Analysis' is useful and can be linked in with site meetings. Update results of the client's network, if properly summarized and distributed, provide the much-needed co-ordination of design and administrative effort. Short-duration schedules are adequate.

(*iii*) *Communications* – Supervision is extremely difficult to maintain. The tying-in of many contractors having varying degrees of planning efficiency is often difficult. Where new contractual procedures are in use they should be depicted in detail as an aid to communication between client and contractor. The contractor's head office plan must be acceptable to site.

(*iv*) *Organization* – The effectiveness of the technique is very dependent on the calibre and training of management and client's administrative staff. The network office at site should cover subcontract work. Contractual demarcations create special problems. CPA often forms the backbone of interaction between various subcontractors; this is particularly true in relation to work in confined spaces, e.g. service shafts, operating theatres, etc.

(*v*) *Resource Allocation* – Resource priorities are important; therefore, keeping to schedule may be a more important objective than detailed allocation of resources. Administration resources can be a serious limitation. The ladder technique is sometimes helpful, with a single room as the unit.

(*vi*) *Education and Training* – The main need may be to improve communications by training courses. Low calibre and wide dispersal of staff make good training very protracted. Strong resistance from 'creative' designers is common.

(*vii*) *Cost Control* – Conflicts arise between client and contractor because their cost aims are different. Penalty/bonus clauses are attractive but are often invalidated. Detail cost control with networks is feasible and desirable but each company may need special computer programs. 'Rate of work' on activities giving trends in productivity is a useful basis for control. The value to the client of timely completion is not always worked out in cash terms. The value of networks for cost control is restricted because of the primitive cost control system within public authorities.

Repetitive building construction

Typical examples of this type of construction are housing estates and local authority administration.

(*i*) *Network Preparation* – The client's network can co-ordinate many different committees, departments, and others. Contractor unit planning is simple, but integration can be complex despite repeating patterns.

(*ii*) *Monitoring* – Line-of-balance techniques are probably more effective for 'finishings'. The rate of production is sometimes monitored by direct unit sales and can be linked in with site meetings. Update results of client's network, if properly summarized and distributed, provide the much-needed co-ordination of design and administrative effort. Short-duration schedules are adequate.

(*iii*) *Communication* – Supervision is extremely difficult to maintain. The tying-in of many contractors, having varying degrees of planning efficiency, is often difficult. Where new contractual procedures are in use they should be depicted in detail as an aid to communication between client and contractor. The contractor's head office plan must be acceptable to site. When the central office is located away from site it is essential for the network analyst to visit the site frequently.

(*iv*) *Organization* – The effectiveness of the technique is very dependent on the calibre and training of management and client's administrative staff. The network office at site should cover subcontract work. Contractual demarcations create special problems. CPA often forms the backbone of interaction between various subcontractors. There are problems in organization with large numbers of small contractors.

(*v*) *Resource Allocation* – Contractors' resources often impose a queueing effect not shown adequately on combined networks. The locality of work may impose severe limitations (this is particularly true in relation to work in confined spaces). There is some overlap with production-control methods. Analysis of resource allocation needs often gives rise to changed logic. The client's administration resources can be a serious limitation. The ladder technique is sometimes helpful.

(*vi*) *Education and Training* – Many subcontractors in building work are not familiar with the technique. Local supervision has to be well trained. Resistance by design groups needs to be overcome. Solicitors and public health inspectors need convincing.

(*vii*) *Cost Control* – Conflicts arise between client and contractor because their cost aims are different. Penalty/bonus clauses are attractive but are often invalidated. Detail cost control with networks is feasible and desirable but each company may need special computer programs. 'Rate of work' on activities giving trends in productivity is a useful basis for control. The value to the client of timely completion is not always worked out in cash terms. The value of networks for cost control is restricted because of primitive cost control systems within public authorities.

Major overhauls

Typical examples of overhauls are blast furnaces, reactor plants, and ship refits.

(*i*) *Network Preparation* – Standards for each contingency help with the difficulty of forecasting all the work necessary. The network needs to allow for rapid shop-floor adjustment.

(*ii*) *Monitoring* – A daily update with job cards is often essential, but problems arise in slick information handling. It is usually preferable to run on the computer only when the changes are significant (i.e. when exceptional corrections must be made). Local staff will put in extra, unscheduled, effort to meet main dates and so maintain the daily programme.

(*iii*) *Communications*—Close contact at shop-floor level is essential. Output data must be easy to understand. A network analyst is only necessary at site on long jobs. Short overhauls (2–3 weeks) can be monitored daily by the planning engineer. Details can be sent by telephone. It is sometimes useful to display the network and mark up progress on bar charts.

(*iv*) *Organization* – A central planning office is essential to build up historical data for the next overhaul. Historical data enable networks to be evolved rapidly and force the engineer to 'think through' his preventive maintenance standards beforehand.

(*v*) *Resource Allocation* – Night-shift working is sometimes involved. It is essential that schedules for trades are realistic. Variable working conditions must be considered. Make more than one resource allocation run to arrive at the best-balanced team. Previous amendments are only a guide, current planning depends on labour availability at the time.

(*vi*) *Education and Training* – The overhaul team should have a clear appreciation of how the technique will schedule and control trades. Low grades are always difficult to train. The briefing of supervision immediately prior to overhaul pays dividends.

(*vii*) *Cost Control* – The value of production compared with the cost of extra effort on overhaul is the important cost factor for control.

12 · Network Preparation

H. WALTON

The initial approach

The initial approach to network preparation depends on the amount of knowledge available about the project. Where the project form and details are obscure or slowly developing, the approach will be via an initial skeleton network. Where, on the other hand, the project is well understood, the detailed network may be drawn straight away.

Networks will be prepared in sections, by geographical or technical areas. These will be developed from skeleton to detailed logic. It is often helpful to draw a master network in which whole subnets are represented by individual arrows. Later, as the detail of the subnets is developed, these arrows can be replaced and interfaced with their appropriate subnets.

Network layout

Networks may be set up in two ways. First, as a large wall-mounted diagram, sometimes even extending to the placing of individual descriptive cards to represent events or activities; second, drawn on a series of sheets of standard size, the display often comprising one subnet per sheet.

The arrangement of the network obviously depends to a very considerable extent on the form of the project. Except on the very largest projects, the layout reflects the technical or geographical areas of the project. These comprise the first level of planning breakdown and bring together all associated activities area by area. Once the logic is expressed by the arrows in each area, it only remains to link or interface these in order to bring the whole network together. Occasionally the project is so large that a higher level of planning breakdown is essential. In such cases, areas of work are so large that they require representation by several subnetworks.

The very large projects are generally non-repeating and are of such a type that the total information on the project is scanty at the beginning. In this situation the concept of a master subnet enables the whole project to be monitored from its initiation. The network, therefore, tends to develop as the project proceeds. Networks relating to the smaller projects can often be built up from almost standard subnets which may be lifted bodily from previous networks. This situation applies particularly in those organizations which tend to repeat the same type of project again and again. The subnet principle is particularly effective in this area as pre-drawn semi-standard subnets can be used and linked together rapidly.

Numbering and referencing

All modern computer programs provide for random event-numbering. True randomness of numbering causes unnecessary difficulty in interpreting the network in that, although it permits less formal event numbering, the event numbers themselves then become non-significant. Particularly on large networks, it is often of considerable use to be able to recognize either sections of the job by groups of numbers or to differentiate between the early part of the project and the later parts by the magnitude of the numbers themselves.

The event number itself can be used as a means of signposting its position within the network. It may be built up from a grid reference based on its position on a particular page. Alternatively, the event number may be coded to represent a particular type of event.

When numbering large networks, particularly when random numbers are used, it is essential to set up some well-established drill to obviate the possibility of using any individual event number more than once. The same numbers may be repeated, however, provided they are separately identified by reference to their individual subnetworks. The standard print-out in i, j order may be used as a tally of event numbers. Alternatively, a separate list may be maintained by marking off used numbers on a standard sheet. For future reference it is useful to include against the issue number of the drawing a statement as to the last event number used.

Computer programs should provide for mnemonic identification of subnets wherever possible, as this assists in the rapid recognition of network and print-out details. Subnet identification can also usefully embody account numbers, contract numbers, area, etc.

Just as a special drill must be set up to prevent the accidental repeated use of event numbers, so it is necessary to index all interface events and their associated leaving and entering activities. Similar treatment should also be applied to start and end events in order to ensure correct discipline with regard to the use of these numbers. A list of such events can usefully be included with the network, the events themselves being given numbers which set them apart from the ordinary network event numbers.

When a diagram is drawn on pages that are mounted together in book form, it is obviously essential that these shall be separately identified. The ability to locate any individual activity depends on the combined indexing capacity of the page references, the subnet identifications, and the event numbers. It is good practice to provide with such networks an index of page contents in terms of the subnets displayed on each page, possibly even in the form of a very skeletonized network linking together all subnets.

Target dates

To enable project management to control progress against a large network, it is of considerable help if the computer program in use can take impressed target dates and handle the negative slack or float which may derive therefrom. The positive and negative values of slack which emerge from calculations against impressed target dates can be used by project management as a direct guide to those activities requiring special attention.

Some computer programs list both primary and secondary slack. The primary slack figures are given against all activities and relate to the final impressed target end-dates. Secondary slack is listed by chains against those activities leading into an event with an impressed target date which is in the body of the network.

The negative slack values are unreal to the extent that they may well be calling for activities to be performed at impossible dates. They are still of use, however, in highlighting to management those activities requiring speed-up or change in logic. Difficulties are encountered, however, when management is not ready to accept such changes, or to recognize that a project may be delayed.

Networks exhibiting negative slack cannot be used for resource allocation. It is under these conditions that project management must find ways of eliminating the negative slack before beginning to allocate resources. This may imply the shortening of durations or the changing of logic by re-deployment of resources.

It is virtually a waste of time to compute any network without impressing at least a finish target date on the end of the network. Without this target date much of the information which management needs to control the project is missing. Management is particularly interested in activities exhibiting negative slack. Without an impressed target date, this would not appear at all.

Milestones

Milestones are key events (or key zero-duration activities) within a network. They are used in two ways: first, as a means of hierarchical reporting from one level of management to the next, and, second, as a means of summarizing a network at the same level.

The first use is somewhat artificial in that it is a device which is employed to enable the computer to produce summarized reports automatically. The second method, namely, the nomination and listing of key events or activities, is geared more to the practical needs of local project management.

Milestones or key events may be passive or active. In the passive sense they

Figure 37 Milestone report

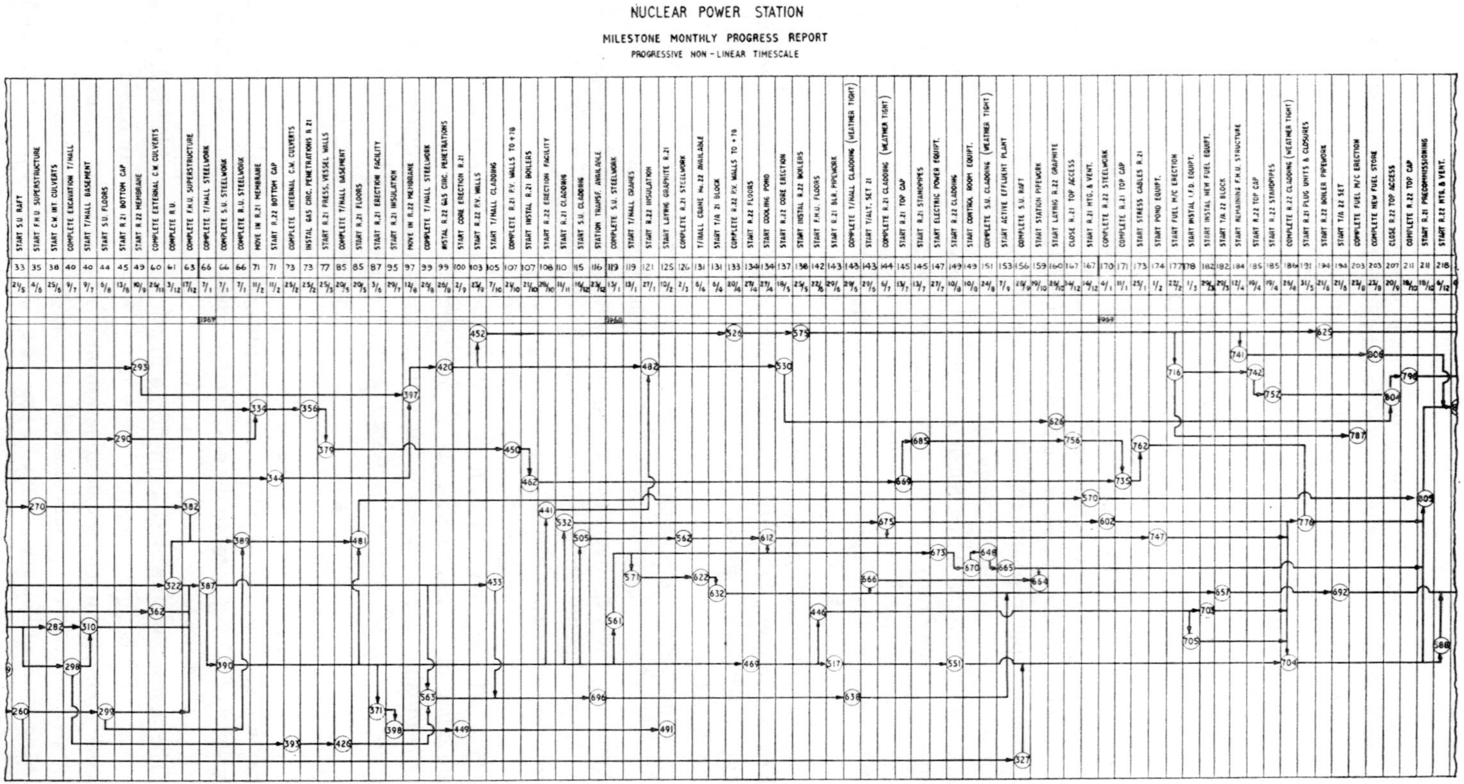

NUCLEAR POWER STATION

MILESTONE MONTHLY PROGRESS REPORT

PROGRESSIVE NON - LINEAR TIMESCALE

Figure 38 *Management presentation of PERT predictions*

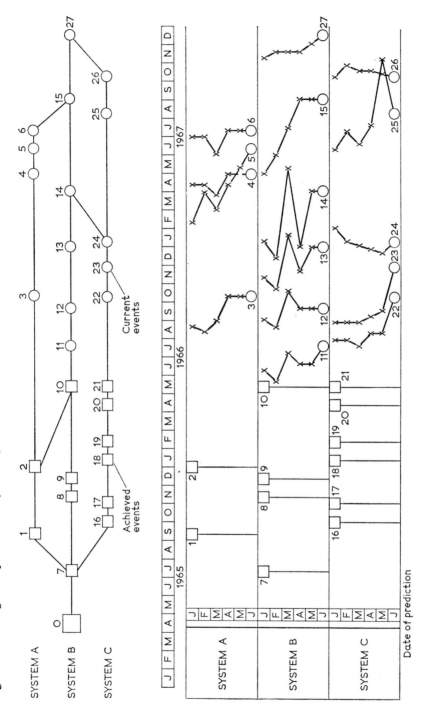

93

are merely special points of interest set up for reporting purposes. In the active sense they carry impressed target dates which have an impact on that part of the network immediately preceding them. Care is required in their application; they should be restricted to a few truly important events.

Figures 37 and *38* show examples of graphical reporting. *Figure 37** plots the network in skeleton form, comprising milestone events only against a non-linear timescale, which merely places the events one after the other. *Figure 38* shows another skeleton network in which the events plot is related to June 1966. Milestone events not yet reached, 3, 4, 11, etc., shown as circles, are plotted at their most recently predicted dates in the skeleton network (upper diagram). The movement over the previous five predictions is shown in the lower diagram.

Grade of staff required – job specification

There are four levels of staff primarily concerned in the preparation of networks; Senior Planners, Planners, Technical Assistants, and Telex/Punch girls.

The Senior Planner should preferably be a graduate with at least five years' relevant experience. He requires to have a knowledge of project technology, and to display initiative and analytical and creative ability. He requires a persuasive personality. He will be a senior man starting in the salary range £2,000 to £2,500 per annum.

The Planner grade should preferably have some qualification and some experience of project technology. He will also require to have analytical ability. His salary may vary widely according to different organizations ranging from £1,100 to £1,900 per annum.

The Technical Assistant should be a technical/clerical man who is able to produce neat and accurate networks and computer input data from information given him by the planners. His salary range will fall somewhere in the bracket £750 to £1,150 per annum. The Telex/Punch girl works direct from the networks and is within the salary range £650 to £750 per annum.

Preparation for stage analysis

It is useful to analyse subnets separately before analysing the network as a whole. The subnet analysis should include an approximate time analysis and revision of sequences, and a computer analysis to eliminate errors of logic and data-punching.

In regard to the elimination of errors, it is often necessary to insert provisional activities to avoid the schedule distortion which would otherwise result from the consideration of one part of the project in isolation. These provisional activities are omitted when the complete network is analysed. In building up the complete network it is useful to link a subnet to the main net-

---

* This method of graphical reporting is no longer used by the CEGB who supplied the diagram for *Figure 37.*—Ed.

work at its start event only for the first computer run and to introduce the additional linkages in stages subsequently. This obviates the possibility of a new subnet with unknown timings upsetting the dates of the main network.

Data-checking

The extent to which computer input data should be checked depends on the cost and availability of computer facilities. Where computer time is hired it is usual to introduce careful checking routines to reduce the cost of processing. Where the user has ready access to his own computer it may be more economical to use successive computer runs to diagnose and eliminate errors. The extent to which error-checking is introduced will, therefore, be decided partly on grounds of computer and staff costs and partly by the efficiency of the diagnostic routines of the computer program, but basically as a result of a detailed economic appraisal.

Data checks include lists of subnet linkages with reference to origin and termination events; checks for duplication of event numbers; checks on number of activities originating at each event ('arrows out' chart); checks for missing information, e.g. durations, or data outside the permitted range; checks for loops; checks for dangling activities. No computer diagnostics will, however, check the accuracy of duration and resource estimates.

13 · Monitoring

D. J. ARMSTRONG

Definition and purpose

Monitoring is the collection of activity progress reports, the refinement of logic, and the re-calculation of the network as a basis for further project control.

Procedure and flow of information

Progress on a project is reported independently by each section. It is not necessary for the entire project to be monitored at a particular instant in time, but this can be carried out to advantage when an organization has its own computer locally.

The ability of the computer to re-sort activities in various ways can be used to assist the monitoring process. The department, or contractor, or even resource manager concerned is coded in the input against each activity. At the output stage, activities are listed not only in chains, say in order of increasing total float, but also by departments, etc., in start date order.

Each department print-out is paged separately so that two copies of each report can be sent to the relevant department, site, or other place as appropriate. The person responsible locally in each department marks up one copy and returns it to the planning officer. Reporting comprises merely a figure against each activity for the time required – zero for completion, and time outstanding for those which have been started. A more reliable return will be obtained by a member of the planning office visiting the site in person, or permanently attached to the site.

Frequency of monitoring

The frequency of monitoring depends on the distance ahead of target dates, the management pattern of the organization, and the availability of a computer. On very long duration projects with far away target dates, it will be sufficient to monitor monthly or even quarterly. On a large network different areas may be monitored at different rates. As the target dates are approached, more and more things become delayed and an increase in monitoring frequency becomes desirable.

Although monitoring may take place at set intervals this does not necessarily mean that the network is to be re-calculated every time. If the reports coming in merely confirm that the set dates for activities are being held, then there is no point in re-calculating.

Monitoring frequency may be very roughly related to the overall duration of the project or the time unit used. A very rough guide for monitoring frequency is 5 per cent of the overall project duration or 20–30 times the time unit. For instance, projects lasting from 1 to 3 years will probably have a time unit of a week and be monitored every month. Projects lasting around 3 months will probably have the day as a time unit and be monitored at weekly intervals. Plant overhauls lasting, say, 14 days will probably have the hour as the time unit and be monitored daily. In practice, however, these figures are extremely variable according to the exigencies of the project. The basic frequency will often be increased when it appears that there are substantial departures from the schedule.

Extent

Full monitoring is usually required on official dates just prior to a project control meeting. Partial or local reappraisal may be conducted either manually or by computer on intermediate dates.

To a limited extent some users are recording actual activity durations, or dates achieved, to provide information for revising estimates of future activities on similar projects.

While only a few users appear to be actively engaged in inter-project scheduling, the monitoring process should be extended to provide information on resources required over a number of projects where they are drawing on the same resource pool.

Management action

The action to be taken as a result of monitoring depends entirely on the management structure of the organization.

Departmental or subcontractor management must inform the project management of new or revised logic; attempt to meet earliest dates; and inform the project management of likelihood of default.

Project management has three main tasks. It must attempt to meet completion dates by realignment of departmental or subcontractor activities, it must inform higher management or the client where delays may be expected, and must seek a policy where appropriate. In projects where circumstances can lead to rapid changes in the network, i.e. with subcontractors on construction sites, it may be necessary to use broadly defined activities. The project manager will take the immediate decisions within the limits of the network.

On very large projects an intermediate programme management function is required to interpret detailed site information for the benefit of project management.

The higher management must ensure that effective action is being taken by project management and must establish policy whenever required.

In the process of monitoring it often happens that the immediate results of calculations are unpalatable. Management at all levels must be educated to react to these figures in the correct way. They must come to realize that there is nothing irrevocable and unchangeable in the results. It is management's job to use the figures and the logic of the network as a basis of analysis to enable them to find ways of bringing a project back on to target.

In many organizations the network analyst is expected to analyse the up-to-date figures manually in order to assess their effect on overall completion. This is done by a quick manual calculation along critical and subcritical paths. If at this stage the results are unacceptable to management, the opportunity can be taken immediately to go back to the offending departments or contractors and endeavour to achieve some speed-up or change of logic to meet the target date. Only when every effort of coercion and of ingenuity in rearranging the logic has been exercised, will management be forced to accept the deteriorating situation and set up a new target date.

14 · Communication

E. G. TRIMBLE

Introduction

Communication means transmission of information. In the case of large, network-controlled projects, the information must be transmitted between different levels of management and operatives, sometimes from one department to another, and always between the planners and the other participants already mentioned. The precise patterns and content of communication must clearly depend on the organization structure. The effectiveness of the communication depends on the ability and willingness of recipients to understand the message. For this reason education and training are required at all levels – see Chapter 16. The reports used as a basis of communication must be carefully designed to suit their individual purpose. The form of reports will be elaborated later.

Control is secured via communication and it is essential that the importance of efficient communication is appreciated at all levels. In particular, it is essential that top management shall not only back the use of network analysis as a control tool, but also understand what is necessary, within the organization, to achieve the vital communications links discussed in this section. Full support is needed from top management in all aspects of network application.

Organization

Many large projects involve a number of different departments. The structure of the organization undertaking a large project may be similar to that shown in *Figure 39*, and in the remainder of this section it is assumed that the organization has this form unless otherwise stated.

Co-ordination will probably be achieved by a small planning section. This section will obtain information from the heads of departments and from individual participants. The results of analyses will be transmitted back to them and to the board (or steering committee).

The geographical location of the planning section is important and the aim should be to establish it at the 'centre of gravity' of the current work-load. This in many projects will move during the execution of the project, for

Figure 39   *Organization structure*

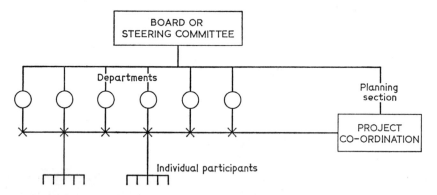

example it may be at head office during the scheme and design stage but on site during the construction stages. In such cases there is a strong argument for changing the location of the planning office during the project. The change would clearly have to be given careful thought; this problem is further considered in Chapter 15.

Reports

The initial collection of information by the planning section will usually be done by informal discussion. Draft networks will be prepared and passed to the participants for comment and adjustment. Estimates of duration and resource requirements will be made, usually on specially prepared and manually completed forms.

The analyses will normally be prepared by computer and the format of the output will depend on its intended user. A useful format for the planning section is a sort by float either on the whole network or divided by subnet. These reports will guide effort when the analysis shows that target dates cannot be achieved without replanning. It is also useful for similar purposes to obtain a primary sort on subnet with a secondary sort on event numbers, since this facilitates the location of each activity in the subnets. In this case the planning staff can readily select the subnet in question and locate the event numbers. On pp. 90–91 above, a grid number system is described.

Output showing calendar dates is also useful to the planning section for comparison with requirements, but an alternative showing week numbers facilitates the adjustments required in replanning. The output for time analysis without resource levelling will usually show the subnet numbers; the activity start and end event numbers; the activity description and duration;

the earliest start and finish dates; the latest start and finish dates; primary and secondary total floats; scheduled dates; and the departmental code.

A convenient sort for use by individual participants is by earliest start date within department when float is principally positive. When a substantial number of activities carry negative float, a sort on latest date is useful since these dates will be earlier than the earliest dates. Opinions differ as to whether latest dates should be disclosed to participants. A compromise, which has been found to work in certain circumstances, is to show float but not the latest dates. It is useful to have a format which leaves space for updating information. In this case two copies are sent to the participant who is asked to complete one copy and return it at each update. It may be necessary for the planning staff to visit participants to assist them in completing update information and to ensure that this work is completed on time.

For use by board (or steering committee) and by heads of departments, it is useful to designate ten to twenty key events in each department and a similar number in the project as a whole. Outlook reports on each key event are then distributed to department heads and the board. It is useful also to convene a meeting of department heads at each update to review the output. Decisions within their own responsibility can then be formulated. Slippage, which cannot be recovered by the exercise of authority at this level, can conveniently be reported to board level by means of a selection of outlook reports which show this slippage, coupled with recommendations as to remedial action. In this way the board's attention is drawn only to exceptional items.

In a long project it is useful to restrict output issued to participants at detail level to a limited forward period of, say, one and a half times the update period. This saves unnecessary printing and, what is more important, eliminates the loss of confidence which can result from the change of forward dates for reasons that may not be fully appreciated by the recipient. There are occasions when a head of department will need a full report in order to assess his forward budget for resources.

Terminology and layout

Communication difficulties arise because of different terminology employed by computer manufacturers in their programs. For example the IBM PERT/COST program uses the terms Expected Start for the earliest start of an activity and Primary Slack for total float (and *not* for event slack); the ICT PERT program uses the terms Early and Late for floats, and has special Ladder activities.

A glossary of CPA standard terms and symbols, which has been submitted to the British Standards Institute, is given in Appendix 1. It is recommended that these terms should be in all CPA programs. Simpler English could improve general acceptability. Words such as 'predecessor' and 'successor' are deprecated.

Detailed attention to layout and printing can save time, temper, and errors. For example, dots are better than noughts in front of event numbers, e.g. . . . 13 is better than 00013. A further suggestion is to print rows in blocks of five.

Negative float

The generation of negative float implies that the earliest date for one or more target dates is too late. Some operators consider that working schedules which show negative float should not be issued to participants. Instead, all the necessary replanning should be done to eliminate the calculated overrun

Figure 40   *Updating cycle*

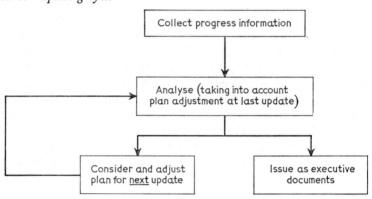

before issuing results of analysis. If this is proved to be impossible the board should be asked to postpone targets to dates which are achievable (see also p. 91).

The delay incurred in replanning can be unacceptable in very complex projects. In such cases a recycling arrangement as shown in *Figure 40* is to be preferred. With this arrangement the acceptance of negative float is essential.

Data transmission

The transmission of data is a continually developing science, and anything written on the subject will quickly be out of date. The stage of development reached in 1967 is the use of data transmission terminals with direct access to an on-line computer. The terminal will be provided with a console feeding the program control instructions to the computer, a line printer and Cathode Ray Tube display to extract limited output from the results. The immediate availability of the selected output will reduce the need for the mass of output provided in the past for checking purposes.

Private GPO lines are now available and capable of providing the high quality transmission of data to ensure reliability, supplemented by parity checking equipment designed into the hardware of the terminals.

Network as a means of communication

In the size of project that warrants a large network, communication between participants has hitherto been limited in accuracy, extent, and logic. The use of network analysis forms a set of precise communication channels and, in large projects, this aspect assumes increasing importance. Top management can receive information which enables control to be exercised specifically instead of by broad generalization as hitherto.

ACKNOWLEDGEMENT

The section on Data Transmission was contributed by Mr R. C. J. Taylor of the CEGB.

15 · Organization

R. C. J. TAYLOR

Organization structures

There are two basic types of organization structure, namely project-oriented management and functional management. Most organizations are variations on these two and hence of a hybrid nature, modifications being necessary to suit each environment. Organization structure is in most cases a restraint for which we have to programme. In a few cases minor organizational changes are possible to accommodate the introduction of a new programme technique. The ideal organizational structure readily available or flexible enough to take full and immediate advantage of the introduction of a new technique is rare indeed.

Two forms of project management systems applied to two different organization structures of extreme types, are explained in detail. A comparison is drawn between the applications of these systems and the organization structure from which they were evolved.

Project management systems

In very large organizations, project-oriented to control a number of large single projects running in parallel at various stages of development, the most recent project management system based on network control is termed 'Joint Planning'. Similarly, another large national organization, functionally oriented, designed an effective measure of project network control, in spite of an unsuitable organization structure, using a 'Steering Committee'. The concepts of these two systems give a lead to modern thinking by embracing control within an existing organization.

It is impossible to over-emphasize the importance of management in any organization. Its influence is paramount. Any organization structure is primarily dependent for its success on the complete support and active participation of management.

Joint Planning

Joint Planning aims to encourage active co-ordination between all contractors and personnel engaged on the project with the contracted intention of completing the project by the programmed date and meeting any schedule dates specified, within agreed costs limits. There are two main contractual requirements; that the contractor co-operates with the client in applying the principles of Joint Planning and that the contractor agrees to produce networks of the detail required by the client.

The organization structure in which Joint Planning was initially evolved is basically project-oriented, but retains the programming as a headquarters service department to all projects (see *Figure 41*). Each project is allocated to a programming team approximately one year before its start, and six to eight years before its major completion. The programming team initially form the basis of the Joint Planning Office. They are responsible for developing the overall philosophy of the programme, and drawing up a master contract-letting programme in sufficient detail to control the early design phases of the project. The first appointment to the programming team is the Project Joint Planning Engineer who is responsible for the organization, planning, programming, and progressing of the project from its embryo state until completion. Together with his programme team at headquarters, he is a permanent member of the Joint Planning Office. Later on, his site progress and planning department would supersede his headquarters team in the Joint Planning Office at site.

The Assistant Programming Engineer (Headquarters), who should be a trained engineer of broad experience, is responsible for the detailed programming of the project within the philosophy developed in conjunction with the Joint Planning Engineer. He also monitors programmes throughout the later design stages of the project when the Joint Planning Office has been established at site. The Assistant Planning and Progress Engineer (Site), who should preferably have had some headquarters experience, is responsible for developing all programmes from the broad network parameters established at headquarters to the day-to-day detailed planning essential at site. The contractor's Joint Planning Engineer is directly responsible to the contractor's Project Manager for all programming and planning, he must be divorced from any other services or responsibilities. It is vital to write into a contract the requirement that 'he must be given sufficient status to have free access to all offices, workshops, and site, and be able to speak for the contractor's organization with quick access to higher management for major decisions'. Some major contractors are developing site planning organizations to absorb the function of cost control, and are using the post of Site Planner as the training-ground for management; this has raised the status of the Joint Planning Engineer.

Initially the Joint Planning Office is established at the design headquarters. During the early phase of the project, contractors' planning engineers pay

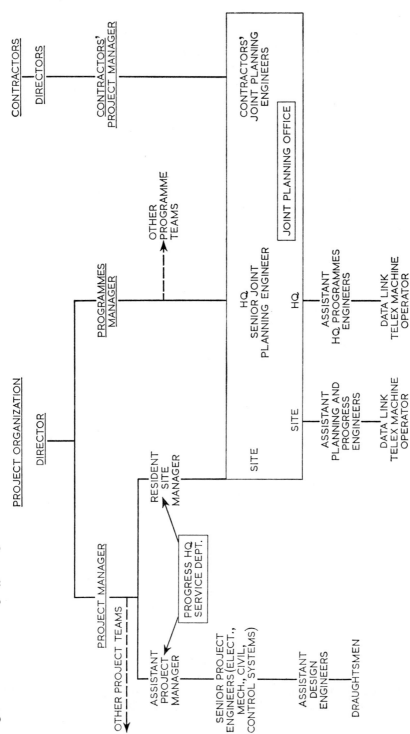

Figure 41 *Joint Planning Office organization tree*

monthly visits to the Joint Planning Office. They obtain a concept of the general philosophy behind Joint Planning and familiarize themselves with established network procedures. They gradually develop detailed sub-networks and help 'debug' these networks for the computer prior to inter-facing these into the master network. The contractor's planning engineer starts to generate the flow of communications between the project and his organization. He sets up monitoring procedures and makes direct contact between project planning engineers and their specialist staff delegated to individual sections of the contract.

As soon as possible during the temporary works period, accommodation should be available to enable the Joint Planning Office to be established at site. Each contractor's Joint Planning Engineer will be a resident member of the Joint Planning Office whenever the contractor is actively engaged on site, and will be required to attend any meeting held which would affect his organization. It is important that each contractor's Joint Planning Engineer spends a considerable amount of time at site, prior to site access, developing detailed erection and commissioning programmes, in conjunction with the site planning team. The Joint Planning Engineer's site team will be responsible for the gradual takeover of all programmes which have been computer-pro-cessed by headquarters through the design stages. These network programmes will have to be monitored and updated regularly. The resultant computer print-out is used for future planning and monitoring. It provides a defined agenda for objective discussion and eliminates subjectivity.

Site Joint Planning should be established as early as possible to satisfy the need on site for a forceful planning section. This section must act with speed and executive authority on any problem or unexpected delay. It should analyse the fault, eliminate it in the future, or at least organize an early-warning system for immediate remedial action. This Joint Planning Office pro-vides a platform to discuss and resolve small programme adjustments between contractors without having to call an official meeting. Judicious use of the com-puter print-out presented departmentally encourages more local analysis. This reduces the number of meetings and, since most meetings are largely unproduc-tive, releases manpower for more important work. Although not eliminating meetings, the improved communication reduces attendances at meetings.

A number of staff problems are created by applying Joint Planning and the philosophy of staff appointments must be carefully thought out. Joint Planning Engineers are attached to a project throughout its duration and must be mobile; this is an obvious training-ground for future management. Assistant Programme Engineers may either be highly qualified engineers with rapid promotion prospects, or specialists in programming remaining in the department to provide continuity. The staff selection procedures must consciously maintain the necessary balance between the 'fast streamers' introducing new ideas and developments, and the specialist programmers providing experience.

There are advantages and disadvantages in maintaining programming as a service department in preference to project 'line' responsibility. The advantages are that it: enables programme teams to stay closely knit and makes results of experiments with new programme aids and techniques immediately available to all teams; provides an objective approach to programming; provides a more effective use of programming manpower by easier mobility of personnel between projects; allows special investigation into the effects of multi-project application; assists in creating a consistent programming policy between projects and stimulates the interchange of information and the creation of new ideas. Possible disadvantages are that: special efforts are required to maintain communication between programmes and project design teams; and that the Project Engineer does not have direct control over the production of programmes, although by delegating this function to the specialist he is only fed the information he has requested and does not have to 'sift' to determine priorities.

A further consideration is that, in the dynamic situation of a programming office, it is necessary to keep abreast of current work in programming and planning techniques and to pursue development in this specialized field. The establishment should always be designed to include a small team to provide this development. Ideally it would be an advantage to have a local Scientific Management or Operational Research Department to collaborate with the programme teams in developing new programme techniques in parallel with normal technique applications, thereby proving their usefulness and 'debugging', etc., without affecting individual project control.

The Steering Committee

An organization established to manage the day-to-day operation of a system leads to a functional management structure, i.e. a number of departmental heads each responsible for their own section of the work (see pp. 101–102 above). Where new projects are infrequent it is not justified to reorganize and create a special project team to co-ordinate the work. In this situation the concept of a 'Steering Committee' was evolved.

A senior representative of each department involved on the project, in addition to representatives of the management consultants/planners, combine to form the Steering Committee. This committee has executive responsibility for the project. The Steering Committee develops the overall philosophy for the whole project to steer departmental decisions required in emergencies. The committee meet at predetermined intervals (say bi-monthly), and most of the subjects for discussion are very much in advance of their start date (up to two years). By this means, conflicts are brought forward early enough for objective policy to be determined rather than being forced to make 'coal-face' decisions on critical items. No emergency meetings of the Steering Committee are called; each individual department makes its own decisions, based on the philosophy evolved for the project.

Day-to-day co-ordination of the project is achieved by a small planning section. This section is required to obtain information from the heads of the various departments and from individual participants. The results of analysis are transmitted back to the Steering Committee as a basis for their future decisions.

The geographical location of the planning section is important. It must be at the centre of gravity of the current work-load. In most large projects the centre will move during the execution of the work, starting at head office during the early stages of the project but migrating to site during the final stages of the job. The change of location must be very clearly defined and carefully thought out.

Staff appointments must be designed to require staff to be flexible in location. An alternative worthy of investigation is that headquarters staff hand their work over to planning staff appointed at site. This alternative has, however, the problem of maintaining continuity and is more easily applied to a line than to a functional organization structure.

Comparison between the Steering Committee and the application of Joint Planning

A combination of both the Steering Committee and Joint Planning lends itself to the situation whereby maintenance teams are required to carry out routine maintenance throughout the year but are faced with the annual project of a major overhaul. The Steering Committee can develop the overall plan for the organization and the Site Joint Planning Office can develop the detail programming on a day-to-day basis.

The similarity between the application of Joint Planning in a basically project-designed line management environment, and the application of a Steering Committee to a functionally oriented management structure, is immediately apparent. Both of these methods have been successfully applied to large organizations in the control of very large networks requiring upwards of 20,000 activities and having an order of detail of £1,000 per activity. Less sophisticated variations of the basic principles can be applied to smaller network situations, but it is important to realize that no standard organization structure is designed to be effective in every environment.

Two basic parameters are available to judge the decision whether to apply either Joint Planning or the Steering Committee. First, the degree of control required to manage the project is decided. Second, the cost involved in controlling the project compared with the cost normally incurred in failing to control it is calculated. In both situations the authority of the controlling body is paramount for a successful application. In the project-oriented organization structure, the project manager must have executive authority for the project. In an organization where individual projects are infrequent and the formation of a project team is not considered practicable, the Steering Committee must have executive authority for the project.

The application of network analysis initially as an aid to planning, quickly points the way to obtaining better control of a project. Organization changes become evident as the method of control is introduced, and during the actual application further organization changes are required for the management control system to realize its full potential. Hence, the organization structure must be flexible and must develop with the advances in project management technique.

16 · Education and Training

Personnel involvement

Large networks usually involve a wide variety of organizations and companies, specialist groups and advisers, contractors and suppliers, as well as direct project staff and work supervisors. The precise team make-up will depend on the individual project, the level of involvement of each person, how the project organization is set up, and how the lines of communication are decided.

Experience indicates that the network system considered as a management co-ordination tool can be easily misunderstood and misapplied. Sometimes the system is erroneously assumed to be a specialist tool for professional planners, work-study men, engineers, or computer staff to use 'on their own'. A realistic network plan that covers all aspects of a project will involve and affect many levels of management and functional interests. If, therefore, a network is drawn for only part of these interests or without close discussion with the responsible people concerned, there is a danger that the control of project execution will be out of balance and insufficient from the client's point of view. For large projects it must be assumed that skilled network planners are needed to work in an integrated way with the managers and engineers concerned, and the latter must fully understand the purpose of the exercise and the way in which the network is to be used.

Successful application of the network system, then, can only follow a clear understanding by management of the principles and methods of planning control the technique requires. An education and training programme must be considered a first step when personnel come together to work on a large project network. This implies education for management not closely involved with detail, to give them a realization of the decision/time/cost implications of the system; and training for those who will be closely involved in the direct use of the system as practitioners or as users of the related data and

115

paperwork. Education is understood to be the imparting of knowledge, and training the imparting of an appropriate skill.

Education programmes

These take the form of short appreciation sessions of about half-day duration which should involve talking specifically about the technique in general terms, as well as the project management problems it raises. Content and length will depend on the types of project the company handles and the range of personnel involved.

Generally a client, or the person/agency that he has appointed to run the project, should have the responsibility for ensuring that a programme of education is mounted as and when necessary. It is assumed that the client is a large organization such as a government department, local authority, major corporation, or industrial/commercial enterprise, in which case some externally available expert or consultant may well have to be employed initially for this purpose. In most cases, however, an 'internal' service or planning department will be created and this should handle subsequent training. Senior people, with good practical experience with the technique, should be used for lecturing in order to carry the necessary weight with top management.

Appreciation sessions can be run with the help of prepared visual aids to enable a lot of ground to be covered in a short period. It is often beneficial to include in these sessions some simple practical work, since this is the best way to create interest and to clarify concepts. In this respect, one group of building industry consultants use a network 'management game', which helps in the familiarization needed and in particular with the decision-taking roles and communications problems that are likely to arise.

Even after initial appreciation sessions, it may well be found necessary to hold follow-up refresher sessions or seminars to restimulate interest, to correct any initial misunderstandings, and to add any new knowledge that is to be applied at, say, a later stage of the project or to new projects coming along. Such follow-ups are important because interest in, and objective attitude to, the use of the system can wane at top level. A special approach may have to be made when different professional people and company units join up for projects, and the level of knowledge, experience, and acceptance is variable. It is thought, however, that, as the technique becomes more widely used and accepted, it will become easier to form a complex project team involving outside contractors and suppliers, since each will be gaining knowledge of the network system as more and more use is currently being made of it.

The type of people who should attend appreciation sessions, in rough priority order, are: directors and senior executive staff; clients' advisory specialists; steering committee members; all senior technical specialists, i.e. architects, and civil, mechanical, electrical, power engineers; purchasing managers; accountants and commercial managers; and senior personnel managers.

Short appreciation sessions may also be necessary for such people as:

designers and draughtsmen; foremen and supervisors; and outside consultants', contractors', and suppliers' staff. The main purpose here will be to waylay any suspicion and concern often surrounding the introduction of a new technique. In some cases, however, proper training courses may be necessary for these people.

It cannot be emphasized enough that this technique involves the 'human element'. Therefore, all education and follow-up methods must be tailored to suit the project and the company conditions and must be handled by responsible and acceptable staff. It cannot be expected that an ideal or fully effective network application will be made at the first attempt. Therefore some careful consideration will have to be given to the choice of the first project to be applied. Furthermore, management must realize that the first application is likely to be experimental and if things don't go exactly right this may not be entirely the fault of the technique itself. Fundamental changes in project management approach, which the network system requires, may take a while to introduce.

When a company is initiating the use of this technique, it is important that good outside lecturers are used and that the main people concerned are encouraged to read all the available literature and to make contacts with other companies more experienced than themselves who are using the technique. Nowadays companies are more prepared to exchange knowledge and experience of a non-competitive kind, so the overall development rate can be that much faster.

Training programmes

Training means a formal approach to the quickest possible development of expertise and experience, which is graded according to the different levels of involvement. If a team is formed for a unique project and afterwards disbanded, or if its involvement in using the technique is intermittent, a specially difficult training problem exists. However, training should not be considered as a 'once-for-all' job. Like many other new management techniques, training is really a continuing process, not only because of the practical experience necessary but also, as in this case, because of the high rate of development of current techniques. Standard approaches will soon be vital.

It is essential that all companies and organizations that are involved in projects shall have at least one of their staff fully knowledgeable and expert in this field. This man's initial training is usually gained from some external source (e.g. a management consultant) and subsequently by direct involvement in the planning and control of actual projects. He should become the source of training and development of other personnel in the use of the technique, and of guidance in setting up the appropriate project team, control procedures, etc. For the larger organizations some project co-ordination department may well be set up with such a man as its head. A suggested outline of a training course is given in the next section.

C P A–I

Except in the very largest organizations, there will not usually be enough practitioners requiring training at any one time to justify a formalized internal course. Such men may, therefore, be sent to external training centres at the first opportunity. Meanwhile, they will receive instruction in the networking department itself as part of the routine application. These courses normally last from two to five days. At present, knowledge of the more advanced techniques, such as resource smoothing, is normally gained from day-to-day experience. In the future, no doubt, this experience will be formalized in separate and more advanced courses. Much depends, however, on a company's use of these techniques and how much it requires people to become involved.

Outline of training course

(A period is one or two days)
Period 1 Introduction to project planning and control concepts.

The fundamental mechanics of networking:

(*i*) Symbols and logic relationships
(*ii*) Diagram layouts
(*iii*) Time estimates and diagram analysis
(*iv*) Slack, float, and the critical path
(*v*) Tabulated analysis for activity start/finish times for scheduling

Progressive exercise that illustrates:

(*i*) Network manipulation
(*ii*) Laddering
(*iii*) Breaking down activities into more detail
(*iv*) Altering activity times and resource requirements
(*v*) Taking account of external restraints
(*vi*) Simple re-allocation of resources

Period 2 Practical networking projects (company-oriented):

(*i*) First with a given list of activities (say 30–50)
(*ii*) Second, up to 100 to 150 activities – preferably without a given list of activities but with a full project situation description

Discussion periods on:

(*i*) Getting networks started in practice
(*ii*) Project organization and the function of network specialists
(*iii*) Level of detail
(*iv*) Master networks and subnetworks, interfacing, etc.

Project control methods

Period 3 Using the computer:

    (*i*) Input
    (*ii*) Types of output
    (*iii*) Programs available

    Resource allocation – principles and methods
    Cost analysis and control – principles and methods
    A general discussion with an 'expert' forum

Period 3 might be considered too much to include in a basic instructional course and it is therefore suggested that this period could be left over until the trainees have had some practical contact (say for a month or so) with the technique. The further period would form a useful refresher at which the extensions of the technique could also be explained depending on whether the company is using these extensively. The discussion session would also be more fruitful and valuable to the people concerned.

ACKNOWLEDGEMENTS

For the most part the Working Party has relied on its own experience and has not in general referred to published works. Additional opinions have, however, been gratefully obtained from J. F. Chilcott (CEGB), F. A. Davies (initially BOC, now Anglo-American), F. A. Moon (CEGB), G. Sneddon (AERE, Harwell), H. S. Woodgate (ICT), V. Wright (CEGB).

In conclusion, the members themselves wish to express their appreciation of the generous attitude of their companies in granting time for these discussions. Acknowledgements are also due to them for the provision of typing facilities and for permission to include certain company diagrams.

Part V

Alternative Approaches to Network Problems

The problems, considered in this collection of papers as critical path or network problems, can also be looked at in other ways. The papers in this chapter show two such ways, Linear Programming and Transportation programming. It is not suggested that these methods of solution are preferable; indeed, the ordinary techniques for solution, because they take advantage of the special features of the problem, are much quicker. However, a mathematical programming approach throws some light on the problems of costs and resources. The first paper shows how the balancing of variable activity costs, briefly mentioned in Chapter 7, can be formulated exactly.

Simple resource allocation problems have been expressed in linear programming form; see, for example, Battersby and Carruthers.[1] Although it is theoretically possible to formulate any resource levelling problem with linear constraints and objective function as a linear program, the labour and difficulty of formulation for a network of significant size would be immense. Moreover a mixed integer programming routine is required and, in spite of recent developments, further improvements in computational efficiency are necessary before this approach becomes a practical proposition. Battersby and Carruthers have also shown that dynamic programming can be used on the resource allocation problem and they have mapped out the lines along which future development of resource techniques might take place.

Apart from Mathematical Programming, various other ways of looking at network problems have been suggested, for example Daniel[2] has drawn an analogy with flows of fluids in pipelines. Alternative methods often throw fresh light on the problem and for special cases may provide useful methods of solution. For example, the Line-of-Balance concept[3] mentioned in Chapter 11, has been successfully used in the construction industry to control large networks which contain subnetworks repeated many times.

REFERENCES

1 A. BATTERSBY and J. A. CARRUTHERS (1966), 'Advances in Critical Path Methods', *Operational Research Quarterly*, **17**, 4.
2 P. T. DANIEL (1965), 'Project Compression: a Technique for Maximising Resource Usage in Closed Loop Systems', Paper to CPASG.
3 N. E. FINK (1965), 'Line of Balance Gives the Answer', *Systems and Procedures Journal*, **9**, 14.

17 · Linear Programming and CPA

S. VAJDA

The critical path

It is our intention, in this paper, to look at critical path scheduling and its associated problems from the point of view of Linear Programming. While this tool is not necessary for the determination of the longest path in a network, it is useful in analysing aspects related to it, as will be seen.

Let a directed network be given, with a start S and a terminal T. It is our first aim to determine the longest chain from S to T, i.e. a succession of

Figure 42   *Simple network*

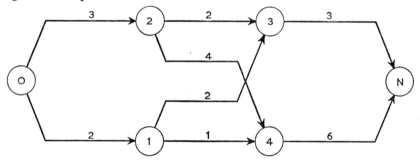

adjacent arcs such that their total length is as large as possible within the network. We denote the nodes by P_i ($i = 0, 1, \ldots, N$), where P_o is the node S and P_N is the node T, and we denote the arcs from P_i to P_j by P_iP_j. The length of P_iP_j will be denoted by c_{ij} and we are looking for the maximum of all expressions.

$$c_{oi_1} + c_{i_1i_2} + \ldots + c_{i_tN}$$

where $(0, i_1), \ldots (i_t, N)$ are pairs of subscripts of nodes P_o and $P_{i_1}, \ldots, P_{i_t}$ and P_N, connected by an arc directed from the first to the second named node. We illustrate our procedure by a typical example, see *Figure 42*.

123

In this case there are four chains P_0 to P_N, indicated by 014N, 013N, 023N, 024N, and the last is the longest of these.

We introduce variables x_{ij} and describe a chain through the network by making $x_{ij} = 1$ if $P_i P_j$ is an arc of the chain, and $x_{ij} = 0$ otherwise. Then, since any chain must start at S and terminate at T, we have

$$\sum_i x_{oi} = 1$$

where the summation extends over those nodes P_i which are directly accessible from S, and we have also

$$\sum_j x_{jN} = 1$$

where the summation extends over those nodes P_j from which T is directly accessible. For reasons which will emerge presently, we prefer to write the latter equation

$$-\sum_j x_{jN} = -1$$

Furthermore, since each node is reached either once, or never, and left if and only if it has been reached at all, we have for each node except S and T

$$-\sum_i x_{ij} + \sum_k x_{jk} = 0$$

where the P_i are nodes from which P_j could be reached directly, and the P_k those which can be reached directly from P_j.

In our example we have thus:

$$
\begin{aligned}
x_{01} + x_{02} &= 1 & -x_{3N} - x_{4N} &= -1 \\
-x_{01} + x_{13} + x_{14} &= 0 & -x_{02} + x_{23} + x_{24} &= 0 \\
-x_{13} - x_{23} + x_{3N} &= 0 & -x_{14} - x_{24} + x_{4N} &= 0
\end{aligned}
\qquad (1)
$$

We note that every variable appears twice, once with a coefficient $+1$ and once with -1, and it is clear that this must be so in every case, not just in this particular example.

We want to find those values x_{ij} which, subject to the given constraints, maximize $\sum_i \sum_j c_{ij} x_{ij}$.

We refer now to a known fact* of the theory of linear programming, viz. that a problem with this pattern of their coefficients can always be written in the form of a transportation problem. We show how this can be done in the present case.

Introduce new variables as follows:

$$
\begin{aligned}
x_{01} + z_1 &= M & x_{13} + x_{23} + z_3 &= M \\
x_{02} + z_2 &= M & x_{14} + x_{24} + z_4 &= M
\end{aligned}
\qquad (2)
$$

where M is a very large number. It will be seen that we have used those combinations of the x_{ij} which appear with negative sign in the constraints.

* Personal communication from George B. Dantzig. This fact is probably known to many, but I cannot remember any explicit statement of it in the literature.

We can then replace every equation which contains positive as well as negative coefficients by two other equations with only positive coefficients, e.g.

$$x_{13} + x_{14} - x_{01} = 0$$

can be replaced by $x_{13} + x_{14} + z_1 = M$, and $x_{01} + z_1 = M$.

By these means the original problem can be written as a transportation
problem as follows:

	1	M	M	M	M
1		x_{01}	x_{02}		
M		z_1		x_{13}	x_{14}
M			z_2	x_{23}	x_{24}
M	x_{3N}			z_3	
M	x_{4N}				z_4

where the x_{ij} have 'cost' c_{ij}, the z_i have cost 0 and all empty cells have a very
large negative cost. Remember, also, that we want to maximize. The reason
for introducing M on the right-hand side of the equation is the same as that
which demands the introduction of large marginal totals in Orden's routine[1]
for solving the trans-shipment problem. If we had not introduced them, then
too many x_{ij} would turn out to be zero.

The solution to this problem is:

	1	M	M	M	M
1		0	1		
M		M			
M			M–1	0	1
M				M	
M	1				M–1

The two 0's were inserted to obtain the requisite number (9) of basic
variables.

To obtain the solution, no elaborate algorithm is required. It can, in fact,
be found by copying any of the well known methods for finding the critical
path, e.g. Roy's Method of Potentials, or the dynamic programming approach, which gives the required longest chain, viz. $P_o P_2 P_4 P_N$. Its length is 13.

Because the optimal solution of a transportation problem is always given in integers if the solution is unique, there is no danger that any x_{ij} would turn out to have some value different from 0 or 1. If an optimal solution with fractional values appears, then this is an indication that there is more than

Figure 43   *Simple network with revised durations*

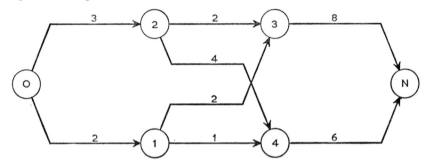

one critical path, and any one of them can easily be recovered. For instance, if the network were as in *Figure 43*, then $x_{o2} = x_{24} = x_{4N} = 1$ and $x_{o2} = x_{23} = x_{3N} = 1$ indicate alternative critical paths, and $x_{o2} = 1$, $x_{24} = x_{23} = x_{3N} = x_{4N} = \frac{1}{2}$ would also be an optimal solution of the transportation problem, but would, of course, not indicate a critical path.

Non-critical paths

The method as here described does not tell us anything about those activities and events which are not on the critical path, and we turn therefore to another procedure, again one of linear programming, to obtain more information.

Let t_i represent the time at which the event indicated by P_i occurs. We make $t_o = 0$. We must then have $t_i \geqslant 0$, and $c_{ij} \leqslant t_j - t_i$ where c_{ij} is the duration of the activity leading from event P_i to P_j. The project is completed at time t_N and the shortest time to completion, i.e. the length of the critical path, is obtained by minimizing t_N, subject to the inequalities given. (The unknowns are $t_1, \ldots, t_N$, while the c_{ij} are known.)

In the example of *Figure 43* we have now:

$$\begin{array}{llll} t_1 \geqslant 2 & t_3 - t_1 \geqslant 2 & t_3 - t_2 \geqslant 2 & t_N - t_3 \geqslant 3 \\ t_2 \geqslant 3 & t_4 - t_1 \geqslant 1 & t_4 - t_2 \geqslant 4 & t_N - t_4 \geqslant 6 \end{array} \tag{3}$$

$$\text{Minimize } t_N.$$

We introduce non-negative slack variables z_{ij} and solve this linear programming problem by any available method.

We then obtain the answer (for instance)

$$t_1 = 2 + z_{o1}$$
$$t_2 = 3 \qquad + z_{o2}$$

$$t_3 = 5 \qquad + z_{02} + z_{23}$$
$$t_4 = 7 \qquad + z_{02} \qquad + z_{24}$$
$$z_{13} = 1 - z_{01} + z_{02} + z_{23} \qquad\qquad (4)$$
$$z_{14} = 4 - z_{01} + z_{02} \qquad + z_{24}$$
$$z_{3N} = 5 \qquad\qquad - z_{23} + z_{24} + z_{4N}$$
$$t_N = 13 \qquad + z_{02} \qquad + z_{24} + z_{4N}$$

The variables on the right-hand side have zero values, and the constants are therefore the values of the variables on the left-hand side.

An increase of z_{02}, z_{24}, or z_{4N} would increase t_N as well, and therefore they must remain at zero. This means that the activities P_oP_2, P_2P_4, and P_4P_N have no float; they indicate the critical path, and its length is 13. On the other hand, z_{01} and/or z_{23} can be increased without altering t_N, so that the solution is not unique.

The signs of the variables on the right-hand side of the expressions for all t_i are positive, so that the values obtained are the earliest for the corresponding events. The alternative solutions can be found by known methods of linear programming, and are as follows (we give here only the values of the basic variables, i.e. of those whose values are not zero, and do not quote their expressions in terms of the other variables):

$$t_1 = 3, \, t_2 = 3, \, t_3 = 5, \, t_4 = 7, \, z_{01} = 1, \, z_{14} = 3, \, z_{3N} = 5, \, t_N = 13$$

or

$$t_1 = 2, \, t_2 = 3, \, t_3 = 10, \, t_4 = 7, \, z_{13} = 6, \, z_{14} = 4, \, z_{23} = 5, \, t_N = 13$$

or

$$t_1 = 6, \, t_2 = 3, \, t_3 = 10, \, t_4 = 7, \, z_{13} = 2, \, z_{01} = 4, \, z_{23} = 5, \, t_N = 13.$$

We can also find the floats from these figures. For instance, for total float, i.e. the interval between the earliest and the latest start of an activity;

$$TF = \max t_j - \min t_i - c_{ij}$$

Free float, i.e. the interval between the earliest finish of an activity and the earliest start of a succeeding activity;

$$FF = \min t_j - \min t_i - c_{ij}$$

Thus we obtain, for our example, the following result:

Activity	Total float	Free float
0–1	4	0
0–2	0	0
1–3	6	1
1–4	4	4
2–3	5	0
2–4	0	0
3–N	5	5
4–N	0	0

We notice that for an activity on the critical path the total as well as the free float is zero, and that for any activity terminating on the critical path the total float equals the free float. This is, of course, as it should be.

Variable activity durations and costs

This linear programming approach has not been mentioned here in order to suggest that it ought to replace more convenient and more efficient methods of computation. However, the linear programming formulation is useful in dealing with more complex problems, involving costs.

We imagine now that the times c_{ij} of the activities are maximum times, which can be reduced by an application of an appropriate amount of money. In particular, assume that if h_{ij} is spent on activity P_iP_j, then its time can be reduced by h_{ij}/e_{ij}. On the other hand, we introduce minimum times, below which a reduction is not possible.

We can then deal with the following problems:

(*i*) Given a certain amount of money, say C, to be expended on reducing the times of the activities, how can we most usefully apply it, i.e. how can we reduce the length of the critical path as much as possible?

(*ii*) If it is desired to reduce the length of the critical path to a given value, say T, how can this be achieved for the smallest expenditure?

We use again our example in *Figure 42* and assume the following values:

$$e_{o1} = 0\cdot5$$
$$e_{o2} = 0\cdot8$$
$$e_{13} = 1\cdot2$$
$$e_{23} = \infty \text{ (i.e. the time of this activity}$$
$$e_{24} = 1\cdot5 \text{ cannot be further reduced)}$$
$$e_{3N} = 0\cdot6$$
$$e_{4N} = 1\cdot2$$

Furthermore, $C = 5$ in problem (*i*), and $T = 10$ in problem (*ii*). The minimum times to which the c_{ij} can be reduced are

$$d_{o1} = d_{o2} = 0, d_{13} = d_{23} = d_{14} = 1, d_{24} = d_{3N} = 2, d_{4N} = 3.$$

We introduce t_{ij} as the time taken for the activity P_iP_j, this is a new unknown. Then the linear programming formulation of the two problems is as follows:

$$
\begin{aligned}
t_{o1} - t_1 + z_{o1} &= 0 & \quad 0 \leqslant t_{o1} \leqslant 2 \\
t_{o2} - t_2 + z_{o2} &= 0 & \quad 0 \leqslant t_{o2} \leqslant 3 \\
t_{13} + t_1 - t_3 + z_{13} &= 0 & \quad 1 \leqslant t_{13} \leqslant 2 \\
t_{14} + t_1 - t_4 + z_{14} &= 0 & \quad 1 \leqslant t_{23} \leqslant 2 \\
t_{23} + t_2 - t_3 + z_{23} &= 0 & \quad 2 \leqslant t_{24} \leqslant 4 \\
t_{24} + t_2 - t_4 + z_{24} &= 0 & \quad 2 \leqslant t_{3N} \leqslant 3
\end{aligned}
\qquad (5)
$$

$$t_{3N} + t_3 - t_N + z_{3N} = 0 \qquad\qquad 4 \leqslant t_{4N} \leqslant 6$$
$$t_{4N} + t_4 - t_N + z_{4N} = 0$$

These constraints are common to problems (*i*) and (*ii*). In addition, in problem (*i*) we have the constraint

$$H' = 0{\cdot}5(2 - t_{o1}) + 0{\cdot}8(3 - t_{o2}) + 1{\cdot}2(2 - t_{13}) + 0{\cdot}6(2 - t_{23})$$
$$+ 1{\cdot}5(4 - t_{24}) + 0{\cdot}6(3 - t_{3N}) + 1{\cdot}2(6 - t_{4N}) \leqslant 5$$

and we want to minimize t_N, while in problem (*ii*) we have $t_N = 10$, and we want to minimize H or, after ignoring the constants,

$$H' = 0{\cdot}5t_{o1} + 0{\cdot}8t_{o2} + 1{\cdot}2t_{13} + 0{\cdot}6t_{23} + 1{\cdot}5t_{24} + 0{\cdot}6t_{3N} + 1{\cdot}2t_{4N}.$$

The answers to these problems are as follows:

(*i*)		(*ii*)
2	t_1	2
0	t_2	0
4	t_3	4
4	t_4	4
47/6	t_N	10
2	t_{o1}	2
0	z_{o1}	0
0	t_{o2}	0
0	z_{o2}	0
2	t_{13}	2
0	z_{13}	0
1	t_{14}	1
1	z_{14}	1
2	t_{23}	2
2	z_{23}	2
4	t_{24}	4
0	z_{24}	0
3	t_{3N}	3
5/6	z_{3N}	3
23/6	t_{4N}	6
0	z_{4N}	0

| For instance (the solution is not unique). | For instance (the solution is not unique). |

These figures mean, that in problem (*i*) we have used 12/5 to reduce t_{o2} from 3 to $3 - 2{\cdot}4/0{\cdot}8 = 0$, and 13/5 to reduce t_{4N} from 6 to $6 - 2{\cdot}6/1{\cdot}2 = 23/6$. This makes the critical path (which is still $P_o P_2 P_4 P_N$) of length $0 + 4 + 23/6 = 47/6$.

In problem (*ii*) it was not necessary to reduce t_N to 47/6, by using $12/5 + 13/5 = 5$. The cheapest way of reducing t_N to 10 was to reduce

t_{o2} to 0, by an application of 12/5. In this case, the minimum value of H was $0.8(3 - 0) = 2.4$, and that of $H' = 0$.

A paper by J. E. Kelley[2] deals with a problem similar to (ii). He takes t_N as a parameter and minimizes a linear function of the t_{ij}, for any given value of t_N. He solves this parametric problem by a primal-dual algorithm.

F. Edmondson and K. B. Haley (in an unpublished communication) consider the problem subject to the same constraints as those above common to problems (i) and (ii), but the objective function $H + c_N t_N$ to be minimized, where H has the same meaning as given earlier, and could, of course, be replaced by H', since this differs from H only by a constant. The problem could be explained to mean that there are overhead expenses which are proportional to the length of time of the whole project, while the times of individual activities can be reduced for an expense. We have then to find the best balance between expenses to reduce individual activities and thereby the total time, and the overhead proportional to the latter.

Taking the same expense rates as above, and $c_N = 1$, this leads, in our example, to the following result:

$$t_1 = 2, t_2 = 0, t_3 = 4, t_4 = 4, t_N = 10$$
$$t_{o1} = 2, z_{o1} = 0, t_{o2} = 0, z_{o2} = 0, t_{13} = 2, z_{13} = 0$$
$$t_{14} = 1, z_{14} = 1, t_{23} = 2, z_{23} = 2, t_{24} = 4, z_{24} = 0$$
$$t_{3N} = 3, z_{3N} = 3, t_{4N} = 6, z_{4N} = 0.$$

Edmondson and Haley have pointed out that, if the number of chains from S to T is small, then the constraints of this problem can be simplified. Thus, in our example, we can write, instead of the equations on the right-hand side of (5)

$$t_N - t_{o2} - t_{23} - t_{3N} \geqslant 0$$
$$t_N - t_{o2} - t_{24} - t_{4N} \geqslant 0$$
$$t_N - t_{o1} - t_{13} - t_{3N} \geqslant 0 \tag{6}$$
$$t_N - t_{o1} - t_{14} - t_{4N} \geqslant 0.$$

The answer is, of course, the same as before as far as the t_{ij} and t_N are concerned, while the slack variables in the last four inequalities are, respectively, 5, 0 (critical path!), 3, and 1.

REFERENCES

1 A. ORDEN (1965), 'The Transshipment Problem', Man. Sci., 2, 276–85.
2 J. E. KELLEY, jr. (1961), 'Critical Path Planning and Scheduling: Mathematical Basis', Ops. Res., 9, 296–320.

18 · A Note on the Transportation Formulation of CPA

J. A. CARRUTHERS

Introduction

The previous chapter indicated that a network could be expressed as a Transportation Problem. This note expands on how this can be done and indicates the way in which slacks and floats can be obtained from this formulation. Let us take the same network as in the previous chapter, see *Figure 44*.

Figure 44 *Simple network*

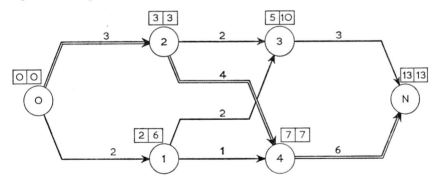

The solution by CPA is analogous to a transportation program for a unit (1) quantity from 0 to N via intermediate destinations to maximize cost, see *Figure 45*. Each node, other than 0 and N, appears twice, as a time inflow and outflow; and, by the identity of time at an event, is connected by a fictitious transport with unit cost zero.

Figure 45   *Transportation diagram*

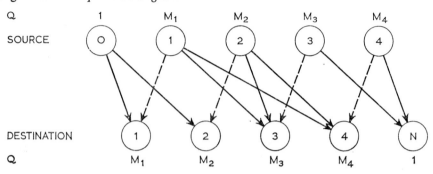

The critical path

For simplicity let the large quantities M_1, M_2, M_3, and M_4 all be equal, M say (there is no need to assume otherwise). Then the transportation matrix can easily be written down and an optimal solution obtained, see *Figure 46*.

Figure 46   *Transportation matrix*

Q	out-going node no.	Q M incoming node no. 1	M 2	M 3	M 4	1 N	Earliest event time
1	0	$X_{o1}^* = 2$	1 $X_{o2}^* = 3$				0
M	1	M $Z_1 = 0$		$X_{13} = 2$	$X_{14} = 1$		2
M	2		$M{-}1$ $Z_2 = 0$	$X_{23}^* = 2$	1 $X_{24}^* = 4$		3
M	3			M $Z_3 = 0$		$X_3 = 3$	5
M	4				$M{-}1$ $Z_4 = 0$	1 $X_4^* = 6$	7

The solution is obtained by inspection, as in Roy's Method of Potentials: Enter 0 as the earliest event time in the top row.

Hence, from row 1 and from $X_{o1} = 2$, enter 2 as earliest time for row two i.e. outgoing node 1.

Hence, for all subsequent rows

$$E_j = \max_{i<j} (E_i + X_{ij})$$

where E_i is earliest time for outgoing node j.

From the last row, event $N - 1$ we have

$$E_n = \max_{i<n} (E_i + X_{in})$$

which is $E_4 + X_{4n} (= 13)$ in our example and this is the project duration. During this process mark those X_{ij} used in the expressions $\max_{i<j} (E_i + X_{ij})$, X_{ij}^* in *Figure 46*. Working back from node N, the continuous path so marked gives us the critical path X_{4n}, X_{24}, X_{02}. Enter 1 as the unit amount for these activities and enter the balance M, or $M - 1$ in the fictitious Z_i cells. These are the basic variables.

Event slacks and activity floats are obtained from the dual values (shadow prices) of rows and columns and from the cell differences, d_{ij}, of non-basic activities. Note that the solution is degenerate, as it will be for all networks which have non-critical events and activities. The indeterminacy of non-critical events can be given limits which are the event slacks.

If the dual values are computed we find they appear for critical events only, thus:

Row	Dual value	Col.	Dual value
0	0	2	−3
2	3	4	−7
4	7	N	−13

These give the event times for critical events; we note that they agree with the earliest times computed by the formula $E_j = \max_{i<j} (E_i + X_{ij})$ while obtaining the critical path. This formula gives earliest times for non-critical activities as well as critical ones. For the non-critical activities enter these earliest times under the corresponding column with a minus sign to complete the set of dual values $(U_i V_j)$. The cell cost differences, $d_{ij} = X_{ij} - C_{ij}$ where C_{ij} are the cell shadow costs, $C_{ij} = -(U_i + V_j)$, can now be computed, and are shown in the top right-hand corner of the cells in *Figure 47* for the non-basic variables.

The latest event times can be obtained by working back from the project duration using the formula $L_i = \min_{j>i} (L_j - X_{ij})$, thus for L_3 we have

$$L_3 = 13 - X_{3N} = 10.$$

Figure 47   *Transportation matrix with event times*

Q out-going node no.	Q M incoming node no. 1	M 2	M 3	M 4	1 N	Dual values $U_i =$ EET	latest event times
1 0	0 1 $X_{o1} = 2$ $X_{o2} = 3$					0	0
M 1	M $Z_1 = 0$		-1 $X_{13} = 2$	-4 $X_{14} = 1$		2	6
M 2		$M-1$ $Z_2 = 0$	0 $X_{23} = 2$	1 $X_{24} = 4$		3	3
M 3			M $Z_3 = 0$		-5 $X_{3N} = 3$	5	10
M 4				$M-1$ $Z_4 = 0$	1 $X_{4N} = 6$	7	7
Dual values V_j	-2	-3	-5	-7	-13		

Slacks and floats

Event slacks are simply the difference between the earliest and latest event times formed above.

Free float, or the leeway on an activity remaining if subsequent activities are started as early as possible, is simply the negative of the d_{ij} calculated above.

Total float can be calculated using the formula total float = free float + end event slack. Other floats can be similarly determined. Thus for our example we obtain the slacks and floats as shown in *Figure 48*.

It is interesting to note that this problem, which superficially has the form of a transportation programming problem, is in fact solved by a non-iterative 'one-pass' method. Since this section was written, Charnes and Raike[1] have drawn attention to a class of generalized network problems which can be solved by 'one-pass' methods.

Figure 48   *Slacks and floats*

Event	Earliest time	Latest time	Slack	Activity	Free float	End event slack	Total float
0	0	0	–	0–1	0	4	4
1	2	6	4	0–2	0	–	–
2	3	3	–	1–3	1	5	6
3	5	10	5	1–4	4	–	4
4	7	7	–	2–3	0	5	5
N	13	13	–	2–4	0	–	–
				3–N	5	–	5
				4–N	0	–	–

REFERENCE

1 A. CHARNES and W. M. RAIKE, 'One Pass Algorithms for some Generalised Network Problems', *Opns. Res.*, **14**, 914–24.

K*

5

Appendix 1

Glossary of Terms and Symbols

The purpose of this glossary is to introduce standard terms, symbols, and definitions to facilitate communication and the spread of knowledge in the subject of project network analysis.

A glossary initially compiled by a working party of the CPA Study Group under the chairmanship of A. G. Simms formed the basis for this glossary prepared by a committee, widely representative of British industry, set up by the British Standards Institution.

Although Project Network Analysis has been available for only a comparatively short time, its use has spread rapidly and its benefits are felt in many fields. In the light of the experience gained, it is now possible to agree on a glossary of terms in common use.

As the whole subject is expanding rapidly, the present glossary is a minimal one, confining itself to essential terms. In future editions more terms will be included, and new sections added to deal with additional methods when these have been applied in practice. Terms and symbols in brackets are permissible alternatives or additions.

General terms

Project network analysis (*Network analysis*)	A group of techniques for presenting information to assist the planning and controlling of projects. The information, usually represented by a *network*, includes the sequence and logical interrelationships of all project *activities*. The group includes techniques for dealing with time, with resources, and with costs.
Critical path analysis	The *project network analysis* technique for determining the minimum project duration.
Activity	An operation or process consuming time and possibly other resources.
Duration	The estimated or actual time required to complete an *activity*.
Path	An unbroken sequence of *activities*.

137

Event	A state in the progress of a project after the completion of all preceding *activities* but before the start of any succeeding *activity*.
Preceding event	An *event* at which an *activity* starts.
Succeeding event	An *event* at which an *activity* finishes.
Start event	An *event* with succeeding but no preceding *activities*.
End event	An *event* with preceding but no succeeding *activities*.
Key event (*milestone*)	An *event* selected for its importance in the project.
Network	A diagram representing the *activities* and *events* of a project, their sequence and interrelationships.
Subnetwork	A *network* which is part of a larger *network*.
Skeleton network	A summary *network*.

Symbols & terms applicable to CPA

Activity	⟶	The arrow represents a logical relationship. Its direction and length need have no significance in representing the *duration*.
Dummy	⇢	A logical link, a constraint which represents no specific operation.
Event	◯	If there is an *event* label it shall be inserted in the circle.
Interface	◎	An *event* which occurs identically in two or more *networks* or *subnetworks*.
Imposed date (*imposed time*)	▽	A date (or point in time) determined by authority or circumstances outside the *network*, or the fixed point for the time scale of the *network*. To be inserted immediately above the *event* concerned.
Earliest date of event (*earliest time of event*)		The earliest date (or point in time) an *event* can occur.
Latest date of event (*latest time of event*)		The latest date (or point in time) an *event* can occur.
		NOTE The *earliest date* shall be placed immediately to the left of (or immediately above) the *latest date* conveniently near the *event* to which they refer on the *network*.
Slack		*Latest date of event* minus *earliest date of event* (may be negative). The term *slack* shall be used as referring only to an *event*.
Earliest start/finish date (*time*) *of activity*		The date (or point in time) before which an *activity* cannot be started/finished.

Latest start/finish date (time) of activity	The date (or point in time) after which an *activity* cannot be started/finished.
Float	A time available for an *activity* or *path* in addition to its *duration* (may be negative).
Total float	*Latest start date of activity* minus *earliest start date of activity* (may be negative).
(Early) free float	*Earliest date* for *succeeding event* of activity minus *earliest finish date of activity*.
Independent float	*Earliest date* of *succeeding event* minus *latest date* of *preceding event* minus activity *duration* (if negative, the *independent float* is taken as zero).
Critical path	A path from a *start event* to an *end event*, the total duration of which is not less than that of any other *path* between the same two *events*.
Critical event	An *event* on the *critical path*.
Critical activity (———||——>)	An *activity* (or *dummy*) on the *critical path*.

Terms used in connection with resource allocation

Resource aggregation	The summation of the requirements of each resource, for each time period, calculated according to a common decision rule.
Scheduling	The process of determining *activity start/finish dates*, subject to resource constraints.
Resource-limited scheduling	The *scheduling* of *activities* such that predetermined resource levels are never exceeded, and the project duration is minimized.
Time-limited scheduling	The *scheduling* of *activities* such that the specified project duration is not exceeded, using resources to a predetermined pattern.
Resource smoothing	The *scheduling* of *activities* within the limits of their *total floats*, such that fluctuations in resource requirements are minimized.

Appendix 2

Selective Reading List

This reading list was originally prepared by a subcommittee of the CPA Study Group under the chairmanship of J. A. Carruthers. At its inaugural meeting on the 22 March 1963, the study group suggested selecting about a dozen relatively accessible publications which could be recommended to anyone wishing to acquire a working knowledge of the techniques. The selection would be reviewed as the techniques developed.

The basic principle underlying the list was to select only the minimum number of publications to cover the essentials. Articles publicizing systems specific to particular commercial organizations were excluded in general. Publications are grouped according to their main purpose, and suggestions on how to obtain them are made where appropriate. According to the Operational Research Society Newsletter of February 1964, US Government Publications which stipulate a price on the cover may be obtained from branches of HM Stationery Office.

General review

'Application of a Technique for Research and Development Program Evaluation',
   D. G. MALCOLM, J. H. ROSEBOOM, C. E. CLARK & W. FAZAR. *Operations Research*, 7, 646–69, September–October 1959.
   General description of the first PERT project.
'Pert as an Analytical Aid for Program Planning—its Payoff and Problems', J. W. POCOCK. *Operations Research*, 10, 893–903, November–December 1963.
   A critical review of the entire field of Critical Path Analysis up to mid-1962.
'PERT Cost', DOD and NASA guide, Systems Design (GPO Cat. No. D1.612:P94).
   US National Aeronautics and Space Administration, June 1962.
   A brief non-mathematical specification of the system and terminology standardized by the DOD and NASA for future applications. Good input/output formats.
   Superintendent of Documents, US Govt. Printing Office, Washington, D.C., at 75 cents. (Try HMSO.)
'PERT Guide for Management Use', PERT Co-ordinating Group (DOD/NASA/ AEC/FAA/BOB), June 1963.
   Available from US Govt. Printing Office.

'Advances in Critical Path Methods', A. BATTERSBY & J. A. CARRUTHERS. *Operational Research Quarterly*, **17**, 359–380, December 1966.

Theory

Conventional

'Critical Path Planning and Scheduling', J. E. KELLEY & M. R. WALKER. *Proceedings*, Eastern Joint Computer Conference, Boston, 1–3 December 1959, 160–73.
A very informative and readable paper.
'Critical Path Planning and Scheduling, Mathematical Basis', J. E. KELLEY. *Operations Research*, **9**, 296–320, May–June 1961.
Mathematical derivation of Critical Path, and least cost method of altering project duration.

Variants

'Graphes et Ordonnancements', B. ROY. *Revue Française de R.O.*, **25**, 323–33, 1962.
Outlines a useful event-oriented alternative to CPM/PERT, not using network diagrams.
An English translation prepared by the National Coal Board has, with the kind co-operation of the NCB, M. Roy, and the *Revue Française de R.O.*, been made available for distribution to members of the OR Society CPA Study Group.
'A Non-Computer Approach to the Critical Path Method for the Construction Industry', 2nd Edition, JOHN W. FONDAHL, Dept. of Civil Engineering, Stanford University, California.
A lucid presentation of an event-oriented alternative to PERT/CPM, including crash costing.
Available from the Construction Institute, Stanford University, at $2.00.
'A Generalized Network Approach to the Planning and Scheduling of a Research Project', H. EISNER. *Operations Research*, **10**, 115–25, January–February 1962.
A sketchy treatment of projects with activities of uncertain outcome.
'An Algebra for the Analysis of Generalised Activity Networks', S. E. ELMA-GHRABY. *Management Science*, **10**, 494–514, April 1964.

Publicity

'Helping the Executive to make up his Mind', BOEHM. *Fortune*, April 1962, 128–31.
'The Analytical Approach to Work', *Financial Times*, 11 February 1963.
'Critical Path Analysis', K. G. LOCKYER. *The Times Review of Industry*, December 1962, 16–17.
'Network Planning and Economic Development', B. O. SZUPROWICZ. *New Scientist*, 27 June 1963, 728–30.
Gives illustrations on National Planning.

Bibliography

Literature

'Critical Path Scheduling', VORESS, HOUSER, and MARSH. TID-3568 (Rev. 1), Division of Technical Information, United States Atomic Energy Commission, July 1962.

From Office of Technical Services, Department of Commerce, Washington 25, D.C., at 75 cents. (Try HMSO.)

'PERT and other Network Planning Systems', Library Bibliography No. 63/23, British Institute of Management Library Service, November 1963.

Computer Programs

'Fifteen Key Features of Computer Programmes for CPM and PERT', C. R. PHILLIPS. *The Journal of Industrial Engineering*, **15**, 14–20, January–February 1964.
Comparison of 36 computer programs available in the USA.

Books

Network Analysis for Planning and Scheduling, A. BATTERSBY. London: Macmillan & Co., 1964.
An Introduction to Critical Path Analysis, K. G. LOCKYER. London: Isaac Pitman & Sons, 1964.
Schedule, Cost and Profit Control, R. V. MILLER. New York: McGraw-Hill, 1963.
Heuristic Methods for Allocating Resources, T. L. PASCOE. Ph.D. Thesis, University of Cambridge, 1965.

Appendix 3

Survey of CPA Programs

The number of organizations offering CPA programs and the range of programs available is such that a newcomer to the field can be bewildered as to the content of programs and their suitability for his purpose. The aim of this Appendix, compiled by the British Ship Research Association under the direction of M. H. Chambers, is to present a summary of the CPA computer programs generally available in Britain. Programs can be obtained through service bureaux, from computer manufacturers for the users of their equipment, or from consultants as part of their service to clients.

The information contained in the table was obtained by circulating a questionnaire to all manufacturers and consultants known to be active in the field and issuing a general appeal for information to members of the CPA Study Group. Information is included on all programs stated by questionnaire respondents to be available in October 1967. Although replies were checked for consistency and queries made where applicable, some omissions and inconsistencies may remain. Data on cost of running programs on a bureau basis were collected, but the methods of charging were so diverse that comparisons would only be misleading. Charges and more detailed information can be obtained from the sponsoring companies.

The information presented will inevitably become outdated as programs are improved and new computers introduced. However, these data should be useful for several years as an indication of the type and range of facilities available. A similar survey of programs available in the US was carried out by Phillips in 1964, see Appendix 2.

CPA computer programs available in UK — 1 October 1967

√ = YES   X = NO   — = NOT APPLICABLE

			W.S. ATKINS & PARTNERS	CEIR		DE LA RUE BULL					
COMPUTER			UNIVAC 1107/1108	IBM 1620	CDC 3200	GAMMA 30	GAMMA 10	GE 115	GE 400	GE 400	GE 400
PROGRAM			C.P.A	CPM/RPSM	CPM/RPSM	PERT	CPM	CPM	CPM; MONITOR	ASTRA	POTENTIAL
GENERAL	1	ACTIVITY NETWORKS	√	√	√	√	√	√	√	√	X
		EVENT NETWORKS	√	X	X	X	X	X	X	X	√
	2	MAX. No. ACTIVITIES OR RESTRAINTS	3,000	2,000	8,000	8,000	∞	2,000	2,300	7,000	3,500
	3	MAX. No. EVENTS OR NODES	1,500	1,400	6,000	4,000	∞	2,000	1,500	3,000	3,500
	4	MAX. No. TIME UNITS	520 WKS / 2600 DYS	2000	10 YRS	9,999	9,999	9,999	9,999	1,000	∞
	5	PROBABILITY NETWORKS	X	X	X	X	X	X	X	X	X
	6	1,2 & 3 TIME ESTIMATES	1	1	1	1 OR 3	1	1	1	1	1
	7	SCHEDULED DATES & NEGATIVE FLOAT	√	√	√	X	√	X	X	X	X
	8	AVAILABLE ON SERVICE BUREAU	√	X	√	X	X	X	√	√	√
SUB-NETS	9	MAX. No. SUBNETS	—	—	—	—	—	—	—	—	—
	10	MAX. No. ACTIVITIES PER SUBNET	—	—	—	—	—	—	—	—	—
	11	MAX. No. INTERFACES PER SUBNET	—	—	—	—	—	—	—	—	—
INPUT	12	SPECIFIED CALENDAR	√	√	√	√	√	X	√	X	X
	13	RANDOM NODE Nos.	√	√	√	√	√	X	√	√	√
	14	RANDOM ACTIVITY LISTING	√	√	√	√	√	X	√	X	√
	15	No. CHARACTERS IN ACTIVITY DESCRIPTION	28	28	39	40	42	28	35	VARIABLE	36
	16	MAX. No. STARTS	100	1	1	E	1	1	1	1	E
		MAX. No. FINISHES	100	1	1	E	1	1	1	1	E
PROCESSING	17	DIAGNOSTIC ROUTINE	√	√	√	X	√	√	X	X	X
	18	SECONDARY FLOAT	√	√	√	X	X	√	√	X	√
	19	RESOURCE TOTALLING	√	√	√	X	√	X	X	√	X
	20	RESOURCE LEVELLING	X	√	√	X	X	X	X	√	X
	21	RESOURCE SMOOTHING	X	√	√	X	X	X	X	X	X
	22	MAX. No. RESOURCES / NETWORK	15	26	26	—	120	—	—	199	—
	23	MAX. No. RESOURCES / ACTIVITY	15	4	4	—	1	—	—	10	—
	24	COST CONTROL	X	√	√	X	X	X	√	√	X
OUTPUT	25	SORTS WITHIN SORTS	√	√	√	√	√	X	X	X	X
	26	CALENDAR PRINT-OUT	X	√	√	√	√	X	√	X	X
	27	BAR CHART	X	√	√	X	X	√	X	X	X
	28	RESOURCE HISTOGRAM	√	√	√	X	√	X	X	√	X
	29	MILESTONE REPORT	√	√	√	X	X	X	√	X	X

1/ THE LIMIT ON 'RESTRAINTS' AND 'NODES' APPLIES TO 'ROY' METHOD PROGRAMS

2/ APPROXIMATE LIMITS FOR A 12K MACHINE

3/ APPROXIMATE LIMITS FOR A 16K MACHINE

continued overleaf

A	= ONLY LIMITED BY THE TOTAL NUMBER OF ACTIVITIES ALLOWED		E	= ONLY LIMITED BY THE TOTAL NUMBER OF EVENTS ALLOWED

ELLIOTT — AUTOMATION COMPUTERS LTD						ENGLISH ELECTRIC — LEO MARCONI COMPUTERS LTD						
803	803	803	4100	4100	903	DEUCE	KDF8	KDF8	LEO III	KDF9	SYSTEM 4-30	SYSTEM 4-50
LO.7/CPA	P.P. MARK II	LO.6	PERT 1	PERT 2	PERT	PERT/RESOURCES	PERT	ROY METHOD	PERT	PERT/RESOURCES	PERT/RESOURCES	PERT/RESOURCES
√	√	√	√	√	√	√	√	√	√	√	√	√
√	X	X	X	X	X	X	√	X	√	√	√	√
4,000 [4]	1,200	2,950	4,095	4,095	680	2,045	1,000	1,000	7,000	10,500 [8]	E	E
3,170 [4]	A	A	3,171	3,171	511	2,046	972	972	4,096	9,000	1,000 [9]	4,000 [10]
524,287	[5]	8191	4,095	4,095	511	9,999	30 YRS	30 YRS [7]	30 YRS [7]	3,276 OR 20 YRS	16,000	16,000
X	√	√	X	X	X	X	√	√	√	√	√	√
1 OR 3	1 OR 3	1 OR 3	1 OR 3	1 OR 3	1 OR 3	1	1 OR 3	1 OR 3	1 OR 3	1 OR 3	1 OR 3	1 OR 3
√	√ [6]	X	√ [6]	√ [6]	√ [6]	√	√	√	√	√	√	√
√	√	√	√	√	X	X	X	√	√	√	X	X
—	—	—	—	—	—	—	—	—	—	9,999	—	—
—	—	—	—	—	—	—	—	—	—	2,500	—	—
—	—	—	—	—	—	—	—	—	—	1,000	—	—
X	X	X	X	X	X	X	X	X	√	√	√	√
√	√	√	√	√	√	√	√	√	√	√	√	√
√	√	√	√	√	√	X	√	√	√	√	√	√
0	0	0	0	36	0	32	40	40	40	55	53	53
E	I	I	E	E	E	E	63	63	E	E	E	E
E	I	I	E	E	E	I	63	63	E	E	E	E
√	X	X	√	√	√	√	X	X	√	√	√	√
√	X	X	√	√	X	X	√	√	√	√	√	√
X	X	X	X	X	X	√	X	X	X	√	√	√
X	X	X	X	X	X	√	X	X	X	√	√	√
X	X	X	X	X	X	X	X	X	X	√	√	√
—	—	—	—	—	—	1	—	—	—	56	100	100
—	—	—	—	—	—	1	—	—	—	7	27	27
X	X	X	X	X	X	X	X	X	X	√	√	√
X	X	X	X	√	X	√	√	√	√	√	√	√
X	X	√	√	√	√	X	√	√	√	√	√	√
X	X	X	X	X	X	X	X	X	√	√	√	√
X	X	X	X	X	X	X	X	X	X	√	√	√
X	X	X	X	X	X	X	√	X	√	√	√	√

4/ APPROXIMATE LIMITS FOR A 8K MACHINE
5/ INFORMATION NOT AVAILABLE
6/ ONLY THE END EVENTS MAY BE SCHEDULED. (NOT INTERMEDIATE EVENTS)
7/ THE CALENDAR LIMIT IS DEC 31st 1999
8/ IF SUBNETS ARE USED THE LIMIT BECOMES 24,997,500 ACTIVITIES
9/ THIS LIMIT RELATES TO THE 32K BITE MACHINE
10/ THIS LIMIT RELATES TO THE 64K BITE MACHINE

CPA computer programs available in UK — 1 October 1967 (*continued*)

COMPANY		MAGNET COMPUTER BUREAU LTD		HONEYWELL		I B M						
COMPUTER		HONEYWELL H.800	HONEYWELL H.200	800	200	7094	7094	1401	1440	S/360	S/360	113C
PROGRAM		PERT	PERT	PERT	PERT	GRASP	PERT/COST	TOTNES	PCS	PMS	PCS	PCS
GENERAL 1	ACTIVITY NETWORKS	✓	✓	✓	✓	✓	✓	✓	✓	✓	✓	✓
	EVENT NETWORKS	✓	✓	✓	✓	X	✓	X	X·	✓	X	X
2	MAX. No. ACTIVITIES OR RESTRAINTS	E	∞	E	∞	5,000	75,000	E	2,000	500,000[3]	5,000	2,00
3	MAX. No. EVENTS OR NODES	16,832	∞	16,832	∞	A	A	7,200	A	A	A	A
4	MAX.No. TIME UNITS	9999 WKS[1]	99,999	9999 WKS[1]	99,999	5 YRS.	10 YRS.	16,383	2,048 DAYS	15 YRS.	9 YRS.	8 Y
5	PROBABILITY NETWORKS	LIMITED	X	✓	X	X	X	X	X	X	X	X
6	1,2&3 TIME ESTIMATES	1 OR 3	1 OR 3	1 OR 3	1 OR 3	1	1 OR 3	1	1	1 OR 3	1	1
7	SCHEDULED DATES & NEGATIVE FLOAT	✓	✓	✓	✓	✓	✓	X	✓	✓	✓	✓
8	AVAILABLE ON SERVICE BUREAU	✓	✓	✓	✓	✓	✓	✓	✓	✓	IN 1968	X
SUB-NETS 9	MAX. No. SUBNETS	—	—	—	11	—	100	—	—	254	—	—
10	MAX. No. ACTIVITIES PER SUBNET	—	—	—	∞	—	750	—	—	3,000[14]	—	—
11	MAX. No. INTERFACES PER SUBNET	—	—	—	∞	—	100	—	—	500[14]	—	—
INPUT 12	SPECIFIED CALENDAR	X	✓	X	✓	OPTIONAL	OPTIONAL	✓	✓	✓	✓·	✓
13	RANDOM NODE Nos.	✓	✓	✓	✓	✓	✓	X	✓	✓	✓	✓
14	RANDOM ACTIVITY LISTING	✓	✓	✓	✓	✓	✓	✓	✓	✓	✓	✓
15	No. CHARACTERS IN ACTIVITY DESCRIPTION	35	38	35	38	30	28	25	45	UP TO 99	44	44
16	MAX. No. STARTS	E	∞	2,000	∞	1	9,900	1	1	50,000[15]	1	1
	MAX. No. FINISHES	E	∞	2,000	∞	1	9,900	1	1	50,000[15]	1	1
PROCESSING 17	DIAGNOSTIC ROUTINE	✓	✓	✓	✓	✓	✓	✓	✓	✓	✓	✓
18	SECONDARY FLOAT	X	✓[2]	X	✓	X	✓	✓	✓	✓	X	X
19	RESOURCE TOTALLING	X	✓	X	✓	✓	✓	X	X	✓	✓	✓
20	RESOURCE LEVELLING	X	X	X	✓	✓	X	X	X	X	X	X
21	RESOURCE SMOOTHING	X	X	X	✓	X	X	X	X	X	X	X
22	MAX. No. RESOURCES /NETWORK	—	∞	—	∞	99	100	—	—	∞	100	10
23	MAX. No. RESOURCES /ACTIVITY	—	8	—	8	5	100	—	—	∞	4	4
24	COST CONTROL	X	X	X	X	X	✓	X	✓	✓	✓	✓
OUTPUT 25	SORTS WITHIN SORTS	LIMITED	✓	LIMITED	✓	✓	✓	✓	✓	✓	✓	✓
26	CALENDAR PRINT-OUT	✓	✓	✓	✓	X	✓	✓	✓	✓	✓	✓
27	BAR CHART	X	✓	X	✓	X	✓	✓	✓	✓	✓	✓
28	RESOURCE HISTOGRAM	X	✓	X	✓	✓	X	X	X	✓	✓	✓
29	MILESTONE REPORT	✓	✓	✓	✓	X	✓	X	✓	✓	✓	✓

1/ THE CALENDAR LIMIT IS DEC 1972
2/ NO INDEPENDENT FLOAT
3/ APPROXIMATE FIGURE FOR 128 K BITE MACHINE
4/ THE TOTAL NUMBER OF START, END AND INTERFAC[E] EVENTS AND ACTIVITIES MUST NOT EXCEED 3,500 FOR A 128 K BITE MACHINE
5/ LIMITED ONLY BY THE NUMBER OF SUBNETS USED AND THE RESTRICTION IN NOTE 14

INTERNATIONAL COMPUTERS AND TABULATORS LTD						K & H. BUSINESS CONSLT. LTD	NAT. CASH REGISTER
1500	1301	ORION	SIRIUS	ATLAS	1900 SERIES	IBM 360	315
PERT & PREDICT	PERT	OPUS	CPM	APPRAISE	PERT	CPM/RPSM	PERT
✓	✓	✓	✓	✓	✓	✓	✓
✓	✓	✓	X	X	✓	X	X
6,000	2,450	12,000	1,400	7,000	60,000 [16]	170,000	9900
3,450	2,000	6,000	1,000	5,000	40,950 [16]	100,000	A
9,999	999 WKS	6 YRS.	999	4,096	4,000	10,000	9999·9
X	✓	✓	✓	✓	X	X	✓
1 OR 3	1 OR 3	1,2 OR 3	1	1 OR 3	1 OR 3	1	1 OR 3
✓	✓	✓	X	✓	✓	✓	✓
✓	✓	✓	X	X	✓	✓	X
—	—	—	—	256	500	∞	—
—	—	—	—	A	6,000	170,000	—
—	—	—	—	NO LIMIT	NO LIMIT	—	—
X	✓	X	X	✓	✓	✓	X
✓	✓	✓	X	✓	✓	✓	✓
✓	✓	✓	✓	✓	✓	✓	✓
28	30	56	25	120	240	1,500	30
E	20	127	127	250	E	E	E
E	20	127	127	250	E	E	E
✓	✓	✓	✓	✓	✓	✓	✓
✓	✓	✓	X	X	✓	✓ [17]	X
✓	X	X	X	✓	✓	✓	X
✓	X	X	X	✓	✓	✓	X
✓	X	X	X	X	✓	✓	X
20	—	—	—	128	125	26	—
4	—	—	—	60	60	4	—
X	X	X	X	✓	✓	✓	X
✓	✓	LIMITED	X	✓	✓	✓	X
✓	✓	✓	X	✓	✓	✓	✓
✓	✓	X	X	✓	✓	✓	✓
X	X	X	X	✓	✓	IN PREPARᵗⁿ	X
✓	✓	✓	X	X	✓	✓	X

16/ USING THE SKELETONIZATION FACILITY   17/ ONLY FREE FLOAT

Subject Index

Program Evaluation and Review Technique (PERT), 58, 60, 63, 77, 82–5, 103
  definition, xi, 3
Program methods, 47–8, 54–5
Progress control (*see also* Cost control), 71, 91, 97
Progress reports, 75–6, 103–4
Project control, 28–9, 31, 43
Project duration, estimates of, 59–61
Project management systems, 107–13
  Joint Planning, 108–13
  Steering Committee, 111–13
Project Network Analysis, *see* Network Analysis
Project scheduling (*see also* Scheduling resources), 51
Project selection, 26–8, 35
Public undertakings, on dispersed site, 83
  on localized site, 83–4

'Quanta of responsibility', 32–3

Referencing, 90–1
Repetitive building construction projects, 86–7
Reports on progress, 75–6, 103–4
Rescheduling, 44
Research and development engineering projects, 82–3
Resource accumulation, 45–6, 51, 55
Resource aggregation, definition, 45, 139
Resource allocation, 16, 41, 43–56, 91, 121
  activity output by scheduled start, 49
  in major problem areas, 81
    capital plant construction, 85
    new product introduction, 85
    one-off building construction, 86
    overhaul projects, 88
    public undertakings, 83–4
    repetitive building construction, 87
    research and development engineering, 82
  package programs, 51–5, 143–7
  philosophy of use, 43–5
  program methods, 47–50
  techniques of, 45–7, 51
  terms used in, 139
Resource levelling, 45–7, 51

Resource limitations, 26–7
Resource-limited scheduling, definition, 139
Resource scheduling, 37–9, 51
Resource smoothing, 45–6, 51, 139
Resource totalling, *see* Resource aggregation
Resource utilization (table), 50
Resources remaining (table), 50
Responsibility quanta, 32–3
Restraints, 15–16

Scheduling, definition, 139
Scheduling resources, 37–9, 51, 55, 139
Secondary objectives, 7, 16
Serial Method, 47, 54
Serial-Parallel Method, 47, 54
Skeleton network, definition, 138
Slack, 91, 103, 118
  and Transportation Formulation, 131, 134–5
  definition, 12, 138
Staff training, *see* Education of Personnel
Start event, 18, 20
  definition, 138
Statistical error distribution (time estimates), 58–9
Steering Committee, 107, 111–13
  comparison with Joint Planning, 112–13
Strategy selection, 7, 10–11, 16
Subnetwork (subnet), definition, 138
Succeeding event, definition, 138

Target dates, 91, 97, 104
Terminology, 7–16, 103–4
  glossary of terms used, 137–9
Time analysis, 17–19, 35
Time-limited scheduling, definition, 139
Total float, 103, 127, 134–5
  definition, 13–14, 139
Training (*see also* Education of personnel), 81–7, 115–19
Transportation formulation, 121, 131–5

Updating of information, 33, 55, 72, 104
  by computer, 37, 39

Work packages, cost of, 73–4, 77
Working-day calculations, 14–15, 38, 49

Name Index